Bone and Renal Failure

Contributions to Nephrology

Vol. 64

KARGER

Basel · München · Paris · London · New York · New Delhi · Singapore · Tokyo · Sydney

International Symposium on Bone and Renal Failure,
Antwerp, November 21–22, 1986

Bone and Renal Failure

Volume Editors
Marc E. De Broe, Frank L. Van de Vyver, Antwerp

34 figures and 24 tables, 1988

KARGER

Basel · München · Paris · London · New York · New Delhi · Singapore · Tokyo · Sydney

Contributions to Nephrology

International Symposium on Bone and Renal Failure (1986: Antwerp, Belgium)
Bone and renal failure/International Symposium on Bone and Renal Failure, Antwerp, November 21–22, 1986: volume editors, Marc E. De Broe, Frank L. Van de Vyver.
(Contributions to nephrology; vol. 64)
Includes bibliographies and index.
1. Renal osteodystrophy – Congresses. 2. Chronic kidney failure – Congresses. 3. Aluminum – Toxicology – Congresses. 4. Lead – Toxicology–Congresses. I. De Broe, M.E. (Marc E.) II. Van de Vyver, Frank L. III. Title. IV. Series.
[DNLM: 1. Kidney Failure, Chronic – congresses. 2. Renal Osteodystrophy–congresses.
W1 CO778UN v. 64 / WJ 342 I5926 1986]
RC931.R4I56 1986 616.6'14–dc19
ISBN 3-8055-4738-2

Bibliographic Indices
This publication is listed in bibliographic services, including Current Contents® and Index Medicus.

Drug Dosage
The authors and the publisher have exerted every effort to ensure that drug selection and dosage set forth in this text are in accord with current recommendations and practice at the time of publication. However, in view of ongoing research, changes in government regulations, and the constant flow of information relating to drug therapy and drug reactions, the reader is urged to check the package insert for each drug for any change in indications and dosage and for added warnings and precautions. This is particularly important when the recommended agent is a new and/or infrequently employed drug.

Printed in Switzerland by Werner Druck AG, Basel
ISBN 3-8055-4738-2

Contents

Contr. Nephrol., vol. 64, pp. 1–4 (Karger, Basel 1988)

Histological Manifestations of Abnormal Parathyroid Hormone Metabolism in Renal Osteodystrophy

Steven L. Teitelbaum

Department of Pathology and Laboratory Medicine, Jewish Hospital at Washington University Medical Center, St. Louis, Mo., USA

Renal osteodystrophy is a generic term encompassing the host of biochemical and morphological abnormalities of the skeleton attending renal failure. These morphological changes include osteitis fibrosa, reflecting the effects of hyperparathyroidism and/or osteomalacia or defective mineralization, in a setting of increased, normal or decreased bone mass.

While the general histology of uremic bone disease has been appreciated for some time, it was not until the last two decades that many of the functional correlates of the specific morphological features of renal osteodystrophy were realized. Clearly, such understanding reflects the development of a variety of new techniques, perhaps the most pivotal being the nondecalcified bone biopsy. This tool has enabled bone morphologists to easily discriminate mineralized from nonmineralized (osteoid) bone matrix. Of equal significance was the discovery that tetracyclines bind to newly deposited bone mineral, and in so doing become autofluorescent in nondecalcified sections of bone. This has permitted these antibiotics to serve as histological markers of skeletal kinetics [1]. Hence, a single bone biopsy not only leads to identification of patients with excessive unmineralized matrix, but also permits one to quantitatively determine the kinetics of bone mineralization.

With these tools in band, a number of investigators have utilized the iliac crest bone biopsy to unravel the complexities of renal osteodystrophy. Our particular interest has been the role of deranged metabolism of calcitropic agents, especially parathyroid hormone (PTH) and vitamin D, in the pathogenesis of uremic bone disease. Our initial efforts involved the relationship of PTH to the rate of skeletal remodeling. These studies were prompted by our observations that tetracycline uptake appeared to be enhanced in the uremic akeleton, despite reports that the rate of bone mineralization is suppressed in chronic renal failure [2–4]. To resolve this controversy, we measured, using tetracycline as a kinetic marker, the rates of bone mineralization in hemodialyzed patients whose bone biopsies showed a predominance of either osteitis fibrosa or osteomalacia. These studies yielded two essential pieces of information:

(1) the histological manifestations of hyperparathyroidism on the uremic skeleton (i.e. osteitis fibrosa) are generally predictable by the circulating levels of immunoreactive PTH (iPTH) [5],

(2) the rate of bone mineralization is greatly enhanced in uremic osteitis fibrosa and the magnitude of tetracycline uptake correlates with log iPTH [6].

These studies not only provided insight into the kinetics of hyperparathyroid bone disease, but also raised questions about a component of the uremic bone disease which had received little prior attention, namely hypoparathyroidism. These concerns were prompted by observations in 2 patients with chronic renal failure who had undergone total parathyroidectomy some years ago. Both individuals developed crippling osteomalacia and, despite heroic efforts at treatment with either vitamin D or 25 -hydroxyvitamin D, their skeletal condition did not improve [6]. These findings underscored the importance of PTH in maintaining optimal skeletal health in uremia. Most importantly, following our observations, we and other investigators encountered patients who developed severe osteomalacia following subtotal parathyroidectomy [7], an operation commonly performed on uremic individuals. These observations raised concerns that nonsurgical means of preventing uremic hyperparathyroidism, such as dietary control of phosphorus intake and absorption, may, if rigorously employed, also have deleterious effects on bone. We addressed this troublesome issue by examining bone biopsies taken from 10 dialysis patients with normal blood phosphorus levels and circulating iPTH concentrations less than 10 times elevated [8]. In comparison, we also studied 46 randomly selected dialysis patients.

As anticipated, the iPTH levels in the 'well-controlled' patients were reflected by histological features of lower bone turnover. Specifically, the quantity of peritrabecular fibrous tissue, number of osteoclasts and quantity of osteoid were significantly reduced in this group as compared to the randomly selected dialysis patients. Despite the relative reduction of osteoid, unmineralized bone matrix remained approximately four times increased in our 'low PTH' patients. Most importantly, only 13.6 ± 5.0 (SEM) of the calcification front of biopsies from the 'well-controlled' group assumed a tetracycline label, whereas fluorescence was much more abundant in the random patients (51.4 ± 6.5). Moreover, only 3 of 10 'low PTH' biopsies had double tetracycline labels. Hence, the 'well-controlled' patients did, in fact, have osteomalacia. Furthermore, all 10 'well-controlled' patients had histological evidence of skeletal aluminum intoxication.

The pivotal observations regarding our 'low PTH' group occurred within the ensuing 5 years after study. Two of the 10 patients developed spontaneous rib fractures, 4 experienced diffuse, severe bone pain, and one died. A desferroxamine challenge test resulted in an increase in mean serum aluminium levels from 72 ± 8 to 305 ± 31 μg/l in the 4 individuals with severe bone pain.

These studies underscore the complexity of maintaining optimal levels of circulating PTH in the uremic patient. One can no longer find comfort in realizing that his patients' iPTH levels are 'well-controlled', as we suspect our most compliant patients are more likely to develop clinically significant uremic bone disease than the randomly selected individual on hemodialysis. The most obvious and pertinent issue is, of course, determining optimal parathyroid gland function. Moreover, the relationship of aluminium toxicity to the effects of PTH on bone are incompletely understood. For example, we do not know if the aluminium toxicity occurring in our 'well-controlled' patients reflects hypoparathyroidism leading to low bone turnover which, in turn, promotes deposition of the cation. An alternative, yet plausible, hypothesis would hold that aluminium promotes hypoparathyroidism due to direct parathyroid gland toxicity, leading in turn to the ravages of PTH deprivation on the uremic skeleton. Clearly much needs to be done to unravel the complex roles of aluminium and relative hypoparathyroidism in renal osteodystrophy.

References

1 Fallon, M.D.; Teitelbaum, S.L.: The interpretation of fluorescent tetracycline markers in the diagnosis of metabolic bone disease. Human Pathol. *13:* 416 (1982).

2 Sarnsethsiri, P., Jaworski, Z.F.; Shimizu, A.G.; Frost, H.M.: New bone formation and osteoid seam thickness. Archs Path. *88:* 49 (1969).

3 Pellegrino, E.D.; Blitz, R.M.: The composition of human bone in uremia. Medicine, Baltimore *44:* 397 (1965).
4 Hitt, O.; Jaworski, Z.F.; Shimizu, A.G.; Frost, H.M.: Tissue-level bone formation rates in chronic renal failure, measured by means of tetracycline bone labeling. Can. J. Physiol. Pharmacol. *48:* 824 (1970).
5 Hruska, K.A.; Teitelbaum, S.L.; Kopelman, R.; Richardson, C.A.; Miller, P.; Debnam, J.; Martin, K.; Slatopolsky, E.: The predictability of the histological features of uremic bone disease by non-invasive techniques. Metab. Bone Dis. Rel. Res. *1:* 39 (1978).
6 Teitelbaum, S.L.; Bergfeld, M.A.; Freitag, J.; Hruska, K.A.; Slatopolsky, E.: Do parathyroid hormone and 1,25-dihydroxyvitamin D modulate bone formation in uremia? J. clin. Endocr. Metab. *51:* 247 (1980).
7 Weinstein, R.S.: Decreased mineralization in hemodialysis patients after subtotal parathyroidectomy. Calcif. Tissue Int. *34:* 16 (1982).
8 Delmez, J.A.; Fallon, M.D.; Harter, H.R.; Hruska, K.A.; Slatopolsky, E.; Teitelbaum, S.L.: Does strict phosphorus control precipitate renal osteomalacia? J. clin. Endocr. Metab. *62:* 747 (1986).

Steven L. Teitelbaum, MD, Department of Pathology and Laboratory Medicine,
Jewish Hospital at Washington University Medical Center,
St. Louis, MO 63110 (USA)

Contr. Nephrol., vol. 64, pp. 5–15 (Karger, Basel 1988)

Factors Modifying the Secretion and Action of Parathyroid Hormone in Renal Failure

T.F. Cundy, N.A.T. Hamdy, J.A. Kanis

Department of Human Metabolism and Clinical Biochemistry, The University of Sheffield Medical School, Sheffield, England

The importance of hyperparathyroidism in the genesis of renal osteodystrophy is now accepted, but ideas about the factors which influence the secretion of parathyroid hormone (PTH) and modify its peripheral actions are still evolving. Understanding these facets of parathyroid pathophysiology has considerable relevance to the clinical management of renal osteodystrophy.

Parathyroid Hormone Secretion

The most important stimulus to PTH secretion is a reduction in the serum concentration of ionized calcium. It is now evident that hyperparathyroidism develops early in the course of progressive chronic renal failure and that this is at least in part due to hypocalcemia [1]. The cause of the hypocalcemia has been the source of controversy. The two most popular ideas have been that either phosphate retention [2] or impaired calcemic responses to PTH [3] are responsible. However, these hypotheses are not mutually exclusive [1]. Recently published studies [4, 5] have provided clarification of the mechanisms underlying hypocalcemia and hyperparathyroidism in early renal failure. In patients with mild renal impairment (glomerular filtration rate, GFR, 55–65 ml/min) severe dietary phosphate restriction produced significant increases in serum 1,25-dihydroxy vitamin D (1,25[OH]$_2$D) concentrations, fractional calcium absorption and serum calcium concentrations, together with significant falls in serum immunoreactive PTH concentrations [4, 5]. These changes occurred despite the fact that these patients were hypophosphatemic and had basal serum $1,25(OH)_2D$ levels within the 'normal' range. One interpretation of these data is that intracellular, rather than extracellular, phosphate accumulation impairs $1,25(OH)_2D$ production early in the course of renal failure. Serum $1,25(OH)_2D$ concentrations, whilst normal, in the sense that they are comparable to those seen in normocalcemic, nonuremic subjects, are inappropriately low for the degree of hyperparathyroidism. Thus, a phosphate-induced reduction in the capacity of the kidney to increase $1,25(OH)_2D$ production could have a central role in the genesis of hypocalcemia and hyperparathyroidism. The predilection of the young and of patients with tubulointerstitial forms of renal disease for developing osteodystrophy may therefore be related to increased requirements for $1,25(OH)_2D$ and preferential destruction of the cells synthesizing it, respectively [6].

In dialyzed uremic patients, the incidence and severity of hyperparathyroidism continues to increase despite the maintenance of normocalcemia or even hypercalcemia. In contrast to the many studies that have been undertaken on the etiology of hypocalcemic hyperparathyroidism in early renal failure, this later phase of the disorder has been little studied. The occurrence of hypercalcemic (tertiary) hyperparathyroidism does not necessarily indicate autonomy of parathyroid secretion, since in the majority of such patients acute suppression of PTH secretion occurs in response to calcium infusion [7]. This phenomenon is described as an alteration in 'set-point', i.e. the concentration of serum-ionized calcium that will suppress PTH secretion by a given amount. Cultured parathyroid cells derived from patients with secondary hyperparathyroidism show an alteration in their set-point in vitro. The mechanisms responsible for the changes in set-point have not been well studied. A change in the set-point for PTH release in uremia is associated with an increase in the number of functional cells; thus, an increase in parathyroid gland mass accompanies increased PTH secretion in hemodialyzed patients. This suggests the presence of factors in uremia which stimulate parathyroid gland hyperplasia. Indeed, hyperparathyroidism recurs frequently after parathyroidectomy in patients maintained on long-term hemodialysis [8], a natural history that is quite unlike that which follows parathyroidectomy for primary hyperparathyroidism. Similarly, nonuremic patients who are inadvertently parathyroidectomized during thyroid surgery rarely regain normal parathyroid function, despite the continued hypocalcemic stimulus to growth of the parathyroid gland. In contrast, parathyroid proliferation in hemodialyzed patients may be very aggressive. This is illustrated in studies of heterotopically autografted parathyroid tissue following parathyroidectomy, where invasive, paraneoplastic behavior may be found [9, 10].

Several factors, in addition to calcium, have been shown to influence PTH secretion, although their physiological relevance in uremia is uncertain. Short-term experiments on the effects of 1,25$(OH)_2$D itself upon PTH secretion in experimental systems have given conflicting answers, with either no effect, stimulation or suppression of PTH secretion [11]. Recently it has been demonstrated that 1,25$(OH)_2$D can reduce PTH synthesis by a direct, calcium-independent effect on the genome [12]. Whether the loss of such an effect due to 1,25$(OH)_2$D deficiency in renal failure contributes significantly to hyperparathyroidism in uremia is not known. Our own experience suggests that it does not. Giving large doses of 1,25$(OH)_2$D to a patient with osteodystrophy, failed to suppress PTH when the diet was calcium-deficient [13]. PTH secretion also seems to be responsive to changes in extracellular magnesium concentrations. Increasing serum magnesium suppresses PTH activity in nonuremic subjects [14] and in dialysis patients a reduction in the dialysate magnesium concentration can increase serum PTH [15, 16]. However, the limited range over which serum magnesium concentrations vary in uremia makes this mechanism doubtful. Aluminum also appears to have direct effects on PTH secretion. The parathyroid accumulates aluminum in uremia [16, 17] and short-term exposure to dialysate with high aluminum concentrations acutely suppresses PTH in serum [18, 19]. A state of functional hypoparathyroidism (despite normal or high serum calcium concentrations) is a feature of aluminum-related bone disease [20], and reversal of aluminum loading can be accompanied by resurgent hyperparathyroidism [21].

It has been suggested that β-blocking drugs [22] and the H_2-receptor antagonists, such as cimetidine [23], may lower circulating serum PTH, but no study has convincingly shown any change in parathyroid activity with these drugs. Their major effects may thus be upon the metabolism of immunoreactive, but biologically inert, fragments of PTH [24]. The vitamin D metabolite, 24,25-dihydroxyvitamin D (24,25-$[OH]_2$D) has also been reported to reduce PTH secretion in certain clinical and experimental circumstances [25], and in combination with 1,25$(OH)_2$D has been reported to decrease parathyroid cell mass [26]. Possibly deficiency of this metabolite in uremia is contributing to parathyroid hypersecretion.

Parathyroid Hormone Action

The calcemic action of exogenous PTH is impaired in renal failure, irrespective of the degree of uremia [3, 27]. As indicated above, this phenomenon, which has been attributed to 'skeletal resistance' has been invoked as a cause of the hypocalcemia of renal failure and secondary hyperparathyroidism. Since bone disease is the hallmark of hyperparathyroidism in renal failure, it is difficult to sustain the idea that it arises because there is a resistance of osteoclasts or osteoblasts to PTH. The phenomenon of the impaired calcemic response to exogenous PTH is complex [28], but $1,25(OH)_2D$ can reverse the defect in part [29], suggesting that the latter hormone can indeed modulate PTH-induced calcemic transfer, possibly at quiescent bone surfaces. Notwithstanding, it is unlikely that $1,25(OH)_2D_3$ is essential for the expression of parathyroid bone disease, since in anephric hemodialysis patients, who have virtually no circulating $1,25(OH)_2D$, hyperparathyroidism is the predominant form of bone disease [30]. Further evidence that circulating $1,25(OH)_2D$ does not significantly modulate PTH action on osteoclasts comes from observations on patients treated long-term with 1α-hydroxylated derivatives of vitamin D. In these subjects the relationship between osteoclast numbers on trabecular bone and circulating serum immunoreactive PTH is not altered by therapy [31], suggesting that there has been no change in osteoclast sensitivity to PTH.

Calcitonin

Serum immunoreactive calcitonin (CT) levels are high in chronic renal failure. In part this is due to delayed catabolism of biologically inert CT fragments, but population studies in patients with end-stage renal failure have shown that those with hyperparathyroid bone disease not only have high concentrations of PTH but also lower serum concentrations of immunoreactive CT. Thus, patients with severe parathyroid bone disease have both higher PTH and lower CT values than those without, but there is still a large overlap between these patients groups for both hormones [32]. The use of the PTH/CT ratio discriminates patients with osteitis fibrosa from those without more completely than the use of either measurement alone. On the basis of these observations we have postulated that augmented secretion of CT protected the skeleton against PTH activity [23]. Alternatively, a failure to augment secretion of CT in uremia may result in parathyroid bone disease. Further evidence that CT can modulate the effects of PTH in uremia comes from studies on hemodialysis patients undergoing bilateral nephrectomy. After this procedure parathyroid bone disease resolves transiently, as judged by biochemical and histological criteria [33, 34]. This improvement is associated with a marked rise in plasma CT which persists for several months. The metabolic clearance rate for CT is not further prolonged by nephrectomy, suggesting that increased secretion causes the increase in plasma values. PTH levels do not change after nephrectomy, indicating that osteoclast and osteoblast numbers can be changed by nephrectomy, independently of changes in PTH [33].

Estrogens

Loss of ovarian estrogen production at the menopause is associated with a marked increase in the prevalence of primary hyperparathyroidism, and administration of exogenous estrogens produces a marked reduction in peripheral PTH action, as judged by biochemical markers [35, 36]. Hypothalamic-pituitary gonadal dysfunction is common in chronic renal failure and amongst patients reaching end-stage a disproportionately large number of women have severe parathyroid bone disease [6]. In hemodialysis-treated patients also, women have been reported to have more extensive radiographic changes than men [37], although we have been unable to confirm this using histological criteria. In addition to its well-documented effects on bone, estrogen may also influence the 'set-point' for PTH release [38]. Remarkably little work has been reported on the possible use of sex steroids in uremic patients.

Vitamin A

It has been suggested that vitamin A may also act synergistically with PTH to enhance bone resorption in uremia [39]. Serum vitamin A concentrations are almost invariably high in patients with renal failure, and bone resorption is a feature of true vitamin A toxicity. However, the elevated serum vitamin A in renal failure is related to increased serum concentrations of the transport protein retinol-binding globulin, and neither serum vitamin A concentrations nor the vitamin A/RBP ratio correlate with biochemical or histological evidence of bone resorption [40].

Implications for Treatment

A variety of treatments given in advanced renal failure can affect PTH activity. These include treatments not specifically aimed at reducing PTH secretion, such as dialysis and transplantation; and those specifically intended to reduce PTH secretion, or its activity on target tissue such as the use of phosphate-binders, vitamin D metabolites and surgical parathyroidectomy.

The mainstay of the medical management of hyperparathyroidism is the manipulation of serum calcium concentrations. Regular hemodialysis treatment, by a net infusion of calcium and a net loss of phosphate at each dialysis, tends to correct the hypocalcemia and hyperphosphatemia of end-stage renal failure. This might be expected to reduce PTH secretion and in the short-term it appears to do so [28, 41]. However, the effect is transient and by 12 months resurgent hyperparathyroidism again becomes evident. The prevalence of clinically significant parathyroid bone disease thereafter increases progressively with the duration of maintenance hemodialysis treatment [42]. Curiously, there is as yet little published evidence to demonstrate that there is the same high incidence of parathyroid bone disease in patients on continuous ambulatory peritoneal dialysis (CAPD) [43].

The hazards of aluminum toxicity have encouraged the use of oral phosphate binders other than aluminum hydroxide. Both calcium carbonate [44] and magnesium carbonate [45] have been shown to suppress PTH secretion in the short term, but the long-term effectiveness of such therapy has not been evaluated.

Serum PTH concentrations and the manifestations of parathyroid bone disease can also be suppressed by oral vitamin D therapy. If given in effective dosage the native vitamin, 25OHD, dihydrotachysterol, 1α-OHD_3 and $1,25(OH)_2D_3$ are probably all equally effective. Stopping treatment results in a rapid increase in serum PTH [46], but continued treatment of hemodialysis patients is also associated with resurgent hyperparathyroidism, again despite the maintenance of pre-dialysis normocalcemia [47]. This transient suppression of hyperparathyroidism but later recurrence is similar to the pattern of events seen after the institution of hemodialysis, but hyperparathyroidism is somewhat slower to recur. The short-term efficacy of vitamin D in suppressing hyperparathyroidism is also limited by the degree of hypercalcemia that can be tolerated, since it is the increment in serum calcium that largely determines the degree of PTH suppression [7, 48]. Slatopolsky et al. [49] have attempted to use the direct suppressive effects of $1,25(OH)_2D$ upon PTH production by the use of intravenous, rather than oral, therapy. In uncontrolled studies this approach appeared to be of some benefit, though again the fall induced in serum PTH was proportional to the increment in serum calcium [49].

Most patients show histological signs of hyperparathyroidism by the time end-stage failure is reached, and since there is good evidence that this results initially from impaired $1,25(OH)_2D$ production, attempts have been made, in dialyzed [50] and predialysis [51] patients, to use 1-hydroxylated forms of vitamin D prophylactically. In both circumstances prophylactic therapy appears to delay, rather than prevent, the onset of the disease. In predialysis patients the possibility of accelerating the decline in renal function by producing higher serum calcium is a source of concern.

The possible beneficial effects of $24,25(OH)_2D_3$ on PTH secretion observed in vitro have not been confirmed in longer term studies in vivo, whether this metabolite was given alone or with $1,25(OH)_2D$ [52].

Alternative approaches to the medical management of parathyroid bone disease include the use of agents which antagonise the action of PTH on bone. Early studies using CT were disappointing. Bone disease progressed in the treated patients. Possibly because endogenous CT secretion was suppressed [53]. Pilot studies with second-generation diphosphonates do however suggest that this may still be a promising therapeutic approach. A number of PTH analogues have been developed which have some inhibitory activity in that they compete for PTH receptors but have little agonist activity. It is possible that such compounds could provide a further approach to blocking the target tissue effects of PTH without activating them [54]. The organic thiophosphate WR-2721, originally developed as a radioprotective agent, has also been shown to block some of the peripheral actions of PTH. It also inhibits PTH secretion [55] and has been used in this way in parathyroid cancer [56]. This drug could also have a use in the management of parathyroid bone disease in uremia.

Parathyroid bone disease remains a common and troublesome complication of chronic renal failure. Newer insights into the control of PTH secretion and its actions on bone have permitted a more rational approach to therapy. However, neither prevention nor cure has yet been achieved.

References

1 Massry, S.G.; Ritz, E.: The pathogenesis of secondary hyperparathyroidism of renal failure – is there a controversy? Archs. intern. Med. *138:* 853–856 (1978).

2 Slatopolsky, E.; Calgar, S.; Pennell, J.P.; Taggart, D.B.; Canterbury, J.M.; Reiss, E.; Bricker, N.S.: On the pathogenesis of hyperparathyroidism in chronic experimental renal insufficiency in the dog. J. clin. Invest. *50:* 492–499 (1971).

3 Llach, F.; Massry, S.G.; Singer, F.R.; Kurokawa, K.; Kaye, J.H.; Coburn, J.W.: Skeletal resistance to endogenous parathyroid hormone in patients with early renal failure. A possible cause for secondary hyperparathyroidism. J. clin. Endocr. Metab. *41:* 339–345 (1975).

4 Portale, A.A.; Booth, B.E.; Halloran, B.P.; Morris, R.C.: Effect of dietary phosphorus on circulating concentrations of 1,25-dihydroxyvitamin D and immunoreactive parathyroid hormone in children with moderate renal insufficiency. J. clin. Invest. *73:* 1580–1589 (1984).

5 Llach, F.; Massry, S.G.: On the mechanism of secondary hyperparathyroidism in moderate renal insufficiency. J. clin. Endocr. Metab. *61:* 601–606 (1985).

6 Cundy, T., Hand, D.J.; Kanis, J.A.; Oliver, D.O.; Wright, F.W.; Woods, C.G.: Who gets renal bone disease before beginning dialysis? Br. med. J. *290:* 271–275 (1979).

7 Muirhead, N.; Catto, G.R.D.; Edward, N.; Adami, S.; Manning, R.M.; O'Riordan, J.L.H.: Suppression of secondary hyperparathyroidism in uraemia: acute and chronic studies. Br. med. J. *288:* 177–179 (1984).

8 Cundy, T.; Kanis, J.A.; Woods, C.A.: Recurrent bone disease after parathyroidectomy (submitted for publication).

9 Frei, U.; Klempa, I.; Schneider, M.; Schenermann, E.H.; Koch, K.M.: Tumour-like growth of parathyroid autografts in uraemic patients. Proc. Eur. Dial. Transpl. Ass. *18:* 548–555 (1981).

10 Bernheim, J.; Korzets, Z.; Magen, H.; Krauss, L.; Bernheim, J.: Total parathyroidectomy with autotransplantation in haemodialysed patients with secondary hyperparathyroidism: should it be abandoned? Nephrol. Dial. Transpl. *1:* 122 (1986).

11 Kanis, J.A.; Guilland-Cumming, D.F.; Russell, R.G.G.: in The endocrinology of calcium metabolism, pp. 321–362 (Raven Press, New York 1982).

12 Silver, J.; Naveh-Many, T.; Mayer, H.; Schmelzer, H.J.; Popovtzer, M.M.: Regulation of vitamin D metabolites of parathyroid hormone gene transcription in vivo by the rat. J. clin. Invest. *78:* 1296–1301 (1986).

13 Cundy, T.; Kanis, J.A.; Heynen, G.; Earnshaw, M.; Clemens, T.L.; O'Riordan, J.L.H.; Merrett, A.L.; Compston, J.E.: Failure to heal vitamin D deficient rickets with conventional doses of 1,25 didydroxyvitamin D3. Br. med. J. *284:* 883–885 (1982).

14 Cholst, I.N.; Steinberg, S.F.; Tropper, P.J.; Fox, H.E.; Segre, G.V.; Bilezikian, J.P.: The influence of hypermagnesemia on serum calcium and parathyroid hormone levels in human subjects. New Engl. J. Med. *310:* 1221–1225 (1984).

15 Pletka, P.; Bernstein, D.S.; Hampers, C.L.; Merrill, J.P.; Sherwood, L.M.: Relationship between magnesium and secondary hyperparthyroidism during long-term hemodialysis. Metabolism *23:* 619–624 (1974).

16 Parsons, V.; Papapoulos, S.E.; Weston, M.J.; Tomlinson, S.; O'Riordan, J.L.H.: The long-term effect of lowering dialysate magnesium on circulating parathyroid hormone in patients on regular haemodialysis therapy. Acta endocr., Copenh. *93:* 455–460 (1980).

17 Cann, C.E.; Prussin, S.G.; Gordon, G.S.: Aluminium uptake by the parathyroid glands. J. clin. Endocr. Metab. *49:* 543–545 (1979).

18 Morrissey, J.; Rothstein, M.; Mayor, G.; Slatopolsky, E.: Suppression of parathyroid hormone release by aluminium. Kidney int. *23:* 699–704 (1983).

19 Cannata, J.B.; Briggs, J.D.; Junor, B.J.; Fell, G.S.; Beastad, G.: Effect of acute aluminium overload on calcium and parathyroid hormone metabolism. Lancet *i:* 501–503 (1983).

20 Ott, S.M.; Maloney, N.A.; Coburn, J.W.; Alfrey, A.C.; Sherrard, D.J.: The prevalence of bone aluminium deposition in renal osteodystrophy and its relation to the response to calcitriol therapy. New Engl. J. Med. *307:* 709–713 (1982).

21 O'Hare, J.A.; Murnaghan, D.J.: Evidence of increased parathyroid activity on discontinuation of high-aluminium dialysate in patients undergoing hemodialysis. Am. J. Med. *77:* 229–232 (1984).

22 Caro, J.F.; Besarab, A.; Burke, J.F.; Glennan, J.A.: A possible role for propranolol in the treatment of renal osteodystrophy. Lancet *ii:* 451–454 (1978).

23 Jacobs, A.I.; Lanier, D.; Canterbury, J.; Bourgoignie, J.J.: Reduction by cimetidine of serum parathyroid hormone levels in uremic patients. New Engl. J. Med. *302:* 671–674 (1980).

24 Fiore, C.; Malatino, L.S.; Kanis, J.A.: Effects of cimetidine on parathyroid hormone metabolism. Lancet *i:* 501 (1981).

25 Canterbury, J.M.; Gavellas, G.; Bourgoignie, J.J.; Reiss, E.: Metabolic consequences of oral administration of 24,25-dihydroxy cholecalciferol to uremic dogs. J. clin. Invest. *65:* 571–576 (1980).

26 Henry, H.L.; Taylor A.N,; Norman, A.W.: Effect of vitamin D metabolites on parathyroid gland size. Proc. Fed. Am. Soc. Exp. Biol. *35:* 340A (1976).

27 Massry, S.G.; Coburn, J.W.; Lee, D.B.N.; Jowsey, J.; Kleeman, C.R.: Skeletal resistance to parathyroid hormone in renal failure: study in 105 human subjects. Ann. intern. Med. *78:* 357–364 (1973).

28 Kanis, J.A.; Cundy, T.: Controversies and uncertainties in the pathophysiology of renal bone disease; in Catto, Coburn, Renal osteodystrophy (Nijhoff, Boston 1988).

29 Sommeville, P.J.; Kaye, M.: Resistance to parathyroid hormone in renal failure: role of vitamin D metabolites. Kidney int. *14:* 245–254 (1978).

30 Bordier, P.J.; Tun Chot, S.; Eastwood, J.B.; Fournier, A.; Wardener, H.E. de: Lack of histological evidence of vitamin D abnormality in the bones of anephric patients. Clin. Sci. *44:* 33–41 (1973).

31 Kanis, J.A.; Cundy, T.; Earnshaw, M.; Henderson, R.G.; Heynen, G.; Naik, R.; Rus-

sell, R.G.G.; Smith, R.; Woods, C.G.: Treatment of renal bone disease with 1-hydroxylated derivatives of vitamin D_3. Q. J. Med. *48:* 289–322 (1979).
32 Heynen, G.; Kanis, J.A.; Ledingham, J.G.G.; Oliver, D.O.; Russell, R.G.G.: Evidence that endogenous calcitonin protects against renal bone disease. Lancet *ii:* 1322–1326 (1976).
33 Kanis, J.A.; Earnshaw, M.; Heynen, G.; Ledingham, J.G.G.; Oliver, D.O.; Russell, R.G.G.; Woods, C.G.; Franchimont, P.; Gaspar, S.: Changes in histologic and biochemical indexes of bone turnover after bilateral nephrectomy in patients on hemodialysis. Evidence for a possible role of endogenous calcitonin. New Engl. J. Med. *296:* 1073–1079 (1977).
34 Cundy, T.; Bartlett, M.; Bishop, M.; Earnshaw, M.; Smith, R.; Kanis, J.A.: Plasma hydroxyproline in uremia: relationships with histological and biochemical indices of bone turnover. Metab. Bone Dis. *4:* 297–303 (1983).
35 Gallagher, J.C.; Wilkinson, R.: The effect of ethinyl oestradiol on calcium and phosphorus metabolism of postmenopausal women with primary hyperparathyroidism. Clin. Sci. *45:* 785–802 (1973).
36 Marcus, R.; Madvig, P.; Crim, M.; Pont, A.; Kosek, J.: Conjugated estrogens in the treatment of post-menopausal women with hyperparathyroidism. Ann. intern. Med. *100:* 633–640 (1984).
37 Meema, H.E.; Oreopoulos, D.G.; Uldall, P.R.: The influence of age and sex on bone resorption of secondary hyperparathyroidism. Calcif. Tissue Int. *36:* 25–30 (1984).
38 Selby, P.L.; Peacock, M.: Ethinyloestradiol and norethisterone in the treatment of primary hyperparathyroidism in postmenopausal women. New Engl. J. Med. *314:* 1481–1485 (1986).
39 Farrington, K.; Miller, P.; Varghese, Z.; Baillod, R.A.; Moorhead, J.F.: Vitamin A toxicity and hypercalcaemia in chronic renal failure. Br. med. J. *282:* 1999–2002 (1981).
40 Cundy, T.; Earnshaw, M.; Heynen, G.; Kanis, J.A.: Vitamin A and hyperparathyroid bone disease in uremia. Am. J. clin. Nutr. *38:* 914–920 (1983).
41 Bouillon, R.; Verberckmoes, R.; DeMoor, P.: Influence of dialysate calcium concentration and vitamin D on serum parathyroid hormone during repetitive dialysis. Kidney int. *7:* 422–432 (1975).
42 Neff, M.S.; Eiser, A.R.; Slifkein, R.F.; Baum, M.; Baez, A.; Gupta, S.; Amarga, E.: Patients surviving ten years on hemodialysis. Am. J. Med. *74:* 996–1004 (1983).
43 Cassidy, M.J.D.; Owen, J.P.; Ellis, H.A.; Dewar, J.; Robinson, C.J.; Wilkinson, R.; Ward, M.K.; Kerr, D.N.S.: Renal osteodystrophy and metastatic calcification on long-term continuous ambulatory peritoneal dialysis. Q. Jl Med. *54:* 29–48 (1985).
44 Slatopolsky, E.; Weerts, C.; Lopez-Hilker, S.; Norwood, K.; Zink, M:; Windus, D.; Delmez, J.: Calcium carbonate as a phosphate binder in patients with chronic renal failure undergoing dialysis. New Engl. J. Med. *315:* 157–161 (1986).
45 O'Donovan, R.; Baldwin, D.; Hammer, M.; Moniz, C.; Parsons, V.: Substitution of aluminium salts by magnesium salts in control of dialysis hyperphosphataemia. Lancet *I:* 880–882 (1986).
46 Kanis, J.A.; Earnshaw, M.; Heynen, G.; Ledingham, J.G.G.; Russell, R.G.G.; Smith, R.; Walton, R.J.: Renal osteodystrophy in nondialysed adolescents: long-term treatment with 1 hydroxycholecalciferol. Archs Dis. Chilh. *52:* 473–481 (1977).

47 Sharman, V.L.; Brownjohn, A.M.; Goodwin, F.J.; Heately, W.; Manning, R.M.; O'Riordan, J.L.H.; Papapoulos, S.E.; Marsh, F.P.: Long-term experience of alfacalcidol in renal osteodystrophy. Q. J. Med. *51:* 271–278 (1982).
48 Nielsen, H.E.; Christensen, M.S.; Melsen, F.; Romer, F.K.; Hansen, H.E.: Effect of 1-hydroxyvitamin D_3 on parathyroid function in patients with chronic renal failure. Clin. Endocr. *7:* suppl., pp. 67s–72s (1977).
49 Slatopolsky, E.; Weerts, C.; Thielan, J.; Horst, R.; Harter, H.; Martin, K.J.: Marked suppression of secondary hyperparathyroidism by intravenous administration of 1,25-dihydroxycholecalciferol in uremic patients. J. clin. Invest. *74:* 2136–2143 (1984).
50 Memmos, D.E.; Eastwood, J.B.; Talner, L.B.; Gower, P.E.; Curtis, J.R.; Phillips, M.E.; Carter, G.D.; Alaghband-Zadeh, J.; Roberts, A.O.; Wardener, H.E. de: Double-blind trial of oral 1,25-dihydroxyvitamin D_3 versus placebo in asymptomatic hyperparathyroidism in patients receiving maintenance haemodialysis. Br. med. J. *282:* 1919–1924 (1981).
51 Healy, M.; Malluche, H.H.; Goldstein, D.A.; Singer, F.R.; Massry, S.G.: Effects of long-term therapy with $1,25(OH)_2D_3$ in patients with moderate renal failure. Archs intern. Med. *140:* 1030–1033 (1980).
52 Muirhead, N.; Adami, S.; Sandler, L.M.; Fraser, R.A.; Catto, G.R.D.; Edward, N.; O'Riordan, J.L.H.: Long-term effects of 1,25 dihydroxyvitamin D_3 and 24,25-dihydroxyvitamin D_3 in renal osteodystrophy. Q. J. Med. *51:* 427–444 (1982).
53 Cundy, T.; Kanis, J.A.; Heynen, G.; Paton, S.; Earnshaw, M.; Woods, C.G.: Deterioration of renal bone disease in patients treated with salmon calcitonin. Clin. Endocr. *16:* 29–37 (1982).
54 Horiuchi, N.; Holick, M.F.; Potts, J.T.; Rosenblatt, M.: A parathyroid hormone inhibitor in vivo: design and biological evaluation of a hormone analog. Science *220:* 1053–1055 (1983).
55 Glover, D.; Riley, L.; Carmichael, K.; Spar, B.; Glick, J.; Kligerman, M.; Agus, Z.S.; Slatopolsky, E.; Attie, M.; Goldfarb, J.: Hypocalcemia and inhibition of parathyroid hormone secretion after administration of WR-2721 (a radioprotective and chemoprotective agent). New Engl. J. Med. *309:* 1137–1141 (1983).
56 Hirshel-Scholz, S.; Jung, A.; Fischer, J.A.; Trechsel, U.; Bonjour, J.P.: Suppression of parathyroid secretion after administration of WR-Z721 in a patient with parathyroid carcinoma. Clin. Endocr. *23:* 313–318 (1985).

J.A. Kanis, MD, Department of Human Metabolism and Clinical Biochemistry,
The University of Sheffield Medical School, Beech Hill Road,
Sheffield, S10 2RX (England)

Contr. Nephrol., vol. 64, pp. 16–24 (Karger, Basel 1988)

Parathyroid Hormone Secretion: Perturbations in Chronic Renal Failure

Eduardo Slatopolsky, Silvia Lopez-Hilker, Adriana Dusso, Jeremiah J. Morrissey, Kevin J. Martin

Renal Division, Department of Medicine, Washington University School of Medicine, St. Louis, Mo., USA

Secretion of Parathyroid Hormone

In the last decade investigators have made substantial progress in the understanding of the factors involved in the regulation of parathyroid hormone (PTH) secretion. PTH is a single-chain protein of 84 amino acids [1]. The hormone is initially synthesized in the rough endoplasmic reticulum of the parathyroid chief cells as the 115 amino acid precursor, pre-pro-PTH [2]. This precursor is converted within seconds of synthesis to another hormone precursor, pro-PTH, which contains 90 amino acids [3]. The pro-PTH in turn is converted to PTH 1–84 by proteolytic cleavage at/or in the vicinity of the Golgi region of the cell [3, 4]. The concentration of the calcium ion in the extracellular space is the primary factor that controls the secretion of PTH [5, 6]. Hypocalcemia is the most potent stimulus for the secretion of PTH. Conversely, hypercalcemia suppresses PTH secretion. There is evidence that a variety of agents, other than calcium, may secondarily modify PTH secretion by affecting the adenosine 3′:5′-cyclic phosphate (cAMP) level of the parathyroid cells [7, 8]. Calcium is an inhibitor of the adenylate cyclase enzyme of the parathyroid glands [9, 10]. The mechanism by which an increasing calcium ion in the extracellular fluid exerts its inhibitory effect on hormone secretion is not precisely known, although it is thought to be mediated in part by an increase in the concentration of calcium ions in the cytosol of the parathyroid cells [11, 12]. Several factors may regulate the level of ionized calcium in the cytosol. These include: (a) movement of calcium into intracellular organels, mitochondria, or elements of the endoplasmic reticulum; (b) the extrusion of ionized calcium from the cell into the extracellular fluid by an ATP-dependent calcium pump [13], (c) a Na^{+}-Ca^{++} exchange system located in the plasma membrane [14, 15]. To assess the role of Na^{+}-Ca^{++} exchange on PTH secretion, Rothstein et al. [16] studied the effects of ouabain and harmaline, specific inhibitors of the Na^{+}-K^{+}-ATPase, on dispersed parathyroid cells in vitro. The results suggest that ouabain and harmaline affect PTH secretion by their action on the Na^{+}-K^{+}-ATPase since inhibition of the sodium pump by potassium removal yields similar results. A decrease in the Na^{+}-Ca^{++} exchange either by blocking the sodium pump or removing external sodium will decrease extrusion of calcium from the cells, thus increasing intracellular calcium which in turn will suppress PTH secretion.

Secretion of PTH in Renal Failure

An alteration in the suppression of PTH secretion by calcium has been shown in glands obtained from patients with chronic renal failure [17, 18]. The set point for calcium is defined as the calcium ion concentration causing a 50% decrease in overall hormone secretion. The set point in normal cells is approximately 1.0 m*M* calcium ion while in patients with primary hyperplasia and secondary hyperplasia, the set point was found to be increased to 1.26 m*M* calcium ion [17, 18]. Figure 1 illustrates the alteration in PTH secretion observed in studies with parathyroid glands in vitro obtained from uremic patients. The relative secretion rate of PTH from a normal gland and a gland that has doubled its size is depicted in figure 1a. The individual cells of the enlarged gland have retained both a normal set point for calcium and their normal sharp responsiveness to calcium to suppress secretion. In this case, serum PTH levels will be increased by an amount that is proportional to the increase in tissue mass at any calcium concentration. Since there is a nonsuppressible component to PTH secretion, there will be increased circulating PTH even in severe hypercalcemia. In figure 1b, in addition to an increase in tissue mass, there is also an increase of the set point for suppression from the normal 1.0 to 1.25 m*M* of calcium. The slope of the calcium suppression remains normal. In this situation, the hyperparathyroidism is a product of both the increase in tissue mass and the lack of suppression by calcium in the normal range. The serum PTH levels are then greater than the proportionate increase in tissue mass. In figure 1c there is not only an increase in tissue mass and an increase in calcium set point, but also less

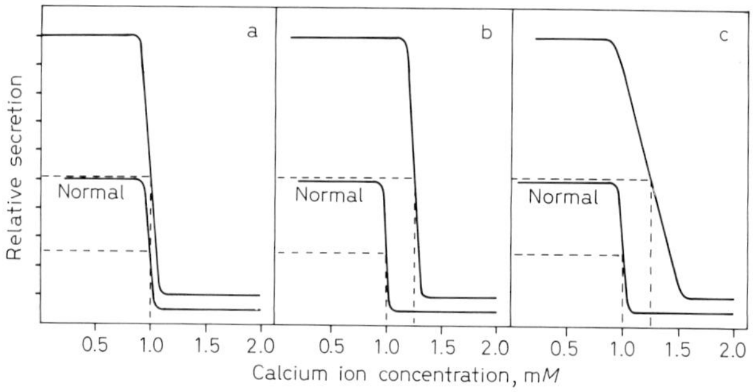

Fig. 1. Diagrammatic representation of factors contributing to abnormal PTH secretion. For description, see text. From Morrissey et al.: Adv. exp. Med. Biol. *178:* 389–398 (1984).

responsiveness of individual cells to calcium. It will seem that the secretion of PTH from hyperplastic glands obtained from uremic patients corresponds to the diagrammatic representation in figure 1c.

The Role of Vitamin D in the Regulation of PTH

Several investigators have provided evidence that vitamin D metabolites directly affect regulation of PTH secretion. In 1974, Oldham et al. [19] isolated a calcium-binding protein from porcine parathyroid glands with properties similar to those of the calcium-binding protein found in mammalian intestinal mucosa. Subsequently, Brumbaugh et al. [20] demonstrated specific binding of $1,25(OH)_2D_3$ to cytosolic and nuclear receptors in the chick parathyroid glands in vitro. Chertow et al. [21] subsequently performed studies in vivo in the rat and in vitro with bovine parathyroid gland slices. These investigators clearly demonstrated an inhibitory effect of $1,25(OH)_2D_3$ on PTH release. Oldham et al. [22] demonstrated in vitamin-D-deficient dogs that higher concentrations of calcium were necessary to suppress the release of PTH in these animals. When similar studies were performed in the same animals after $1,25(OH)_2D_3$ was given, the parathyroid glands appeared to be more sensitive to mild increments in serum calcium. The low levels of $1,25(OH)_2D_3$ observed in patients with advanced renal insufficiency [23, 24] may potentially play a role in the abnormal behavior of the parathyroid glands. Madsen et al. [25] studied the effect of intravenous $1,25(OH)_2D_3$

(250 ng every 6 h) in 10 patients with acute oliguric renal failure. In these patients, serum-ionized calcium was maintained constant and at a subnormal level by continuous peritoneal dialysis. A significant suppression of PTH levels was observed in patients receiving $1,25(OH)_2D_3$. The authors concluded that since the serum calcium was maintained constant by the dialysis procedure, $1,25(OH)_2D_3$ per se suppressed the release of PTH. Recently [26], we performed studies in 20 patients treated by chronic hemodialysis. The patients received intravenous $1,25(OH)_2D_3$ 0.5–4.0 μg/3 times a week. Whereas oral administration of $1,25(OH)_2D_3$ in doses adequate to maintain serum calcium at the upper limit of normal did not alter PTH levels, a marked suppression ($70 \pm 3\%$ of PTH levels) was seen in all 20 patients given intravenous $1,25(OH)_2D_3$. Temporal studies showed a $20 \pm 5\%$ decrease in PTH without significant change in serum calcium with intravenous $1,25(OH)_2D_3$. In 5 patients, the serum calcium was increased by the oral administration of calcium carbonate. The decrement in serum immunoreactive PTH (iPTH) was only $25 \pm 7\%$ when compared with 73% obtained by administration of intravenous $1,25(OH)_2D_3$. Thus, a similar serum calcium achieved by intravenous $1,25(OH)_2D_3$ rather than calcium carbonate has a greater suppressive effect on the release of PTH. These studies indicate that $1,25(OH)_2D_3$ administered intravenously rather than orally may result in a greater delivery of vitamin D metabolite to peripheral target tissues other than the intestine and allow a greater expression of the biological effect of $1,25(OH)_2D_3$ in peripheral tissues, i.e. parathyroid gland. The use of intravenous $1,25(OH)_2D_3$ provides a simple and effective way to suppress secondary hyperparathyroidism in dialysis patients. To study further if $1,25(OH)_2D_3$ has a direct effect on the secretion of PTH in addition to its action on serum calcium, Chan et al. [27] investigated the possibility of a direct suppressive effect of $1,25(OH)_2D_3$ on PTH release in monolayer cultures of bovine parathyroid cells.

Figure 2 illustrates the results of the effects of $1,25(OH)_2D_3$ on carboxyl and amino terminal iPTH secretion after 48 h of incubation. As demonstrated in figure 2 the addition of $1,25(OH)_2D_3$ to the medium has a direct suppressive effect on the release of PTH. Decreased PTH secretion by the cultured cells could be a result of either decrease in PTH synthesis and/or increase in intracellular degradation of the hormone. In fact, a decreased PTH synthesis as reflected by a decrease in the steady-state level of cytoplasmic pre-pro-PTH mRNA in primary parathyroid cell cultures exposed to $1,25(OH)_2D_3$ has been demonstrated [28]. Recent

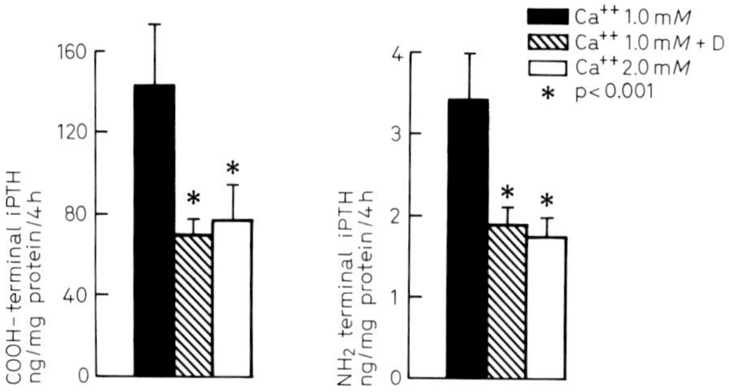

Fig. 2. Effect of $1,25(OH)_2D_3$ on carboxyl and aminoterminal iPTH secretion after 48 h of incubation. Reproduced from Chan et al. [27].

studies [29] in vivo have demonstrated that $1,25(OH)_2D_3$ decreased pre-pro-PTH mRNA over 3–48 h and that $1,25(OH)_2D_3$ directly inhibits gene transcription.

Studies in Dogs with Renal Failure

Since studies with parathyroid cells obtained from uremic patients indicate that there is a shift in the set point for calcium regulated PTH secretion and $1,25(OH)_2D_3$ has a direct suppressive effect on PTH release, Lopez-Hilker et al. [30] performed further studies to determine if in vivo changes in vitamin D metabolism may affect the set point for calcium. Studies were performed in dogs with renal failure. Hypocalcemia was prevented by administration of a high calcium diet. Initially ionized calcium was 4.79 ± 0.09 mg/dl which was gradually increased to 5.3 ± 0.5 mg/dl. Despite a moderate increase in ionized calcium, amino terminal PTH increased from 65 to 118 pg/ml. Serum $1,25(OH)_2D_3$ decreased from 25 to 12 pg/ml. Further studies were performed in 2 other groups of dogs. One group received 150–200 ng and the second 75–100 ng of $1,25(OH)_2D_3$ twice daily. In the second group the level of serum $1,25(OH)_2D_3$ after nephrectomy remained normal and amino terminal iPTH did not increase (fig. 3). Then, in the animals that received a high calcium intake, we observed a decrease in the concentration of $1,25(OH)_2D_3$ immediately after experimental induction of renal insufficiency. However, hyperparathyroidism developed although serum cal-

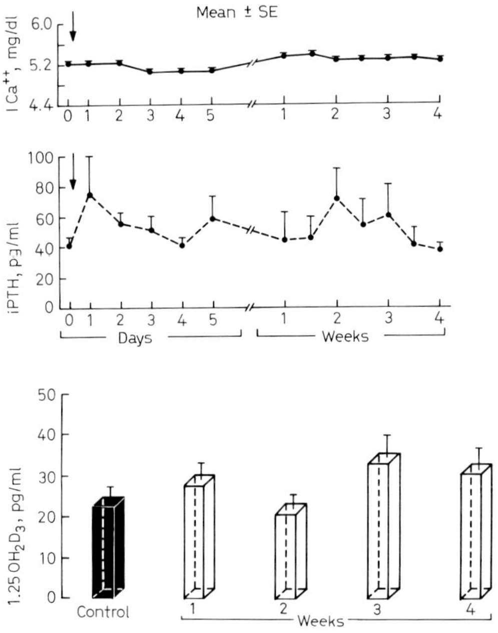

Fig. 3. Serum ionized calcium, amino iPTH levels and serum $1,25(OH)_2D_3$ in a group of 6 dogs who received 75–100 ng of $1,25(OH)_2D_3$ twice daily before and after renal insufficiency (arrow).

cium did not decrease. Thus, the low levels of $1,25(OH)_2D_3$ independent of changes in ionized calcium may have directly affected the secretion of PTH. The absence of an increase of PTH levels in the presence of normal or slightly increased levels of calcium with the administration of $1,25(OH)_2D_3$, supports this hypothesis. Since we avoided a fall in the serum concentration of $1,25(OH)_2D_3$ by its exogenous administration, it is possible that with the early administration of this hormone we prevented the altered synthesis and secretion of PTH associated with low levels of $1,25(OH)_2D_3$. However, we must emphasize that in the group of dogs receiving $1,25(OH)_2D_3$ (150–200 ng twice daily), serum iPTH did not become undetectable despite mild hypercalcemia and high levels of circulating $1,25(OH)_2D_3$. This suggests that in uremic dogs other factors beside calcium and $1,25(OH)_2D_3$ may have a regulating effect on the secretion of PTH.

Since phosphate regulates the conversion of 25(OH)D to $1,25(OH)_2D_3$, it has been postulated that a low phosphate diet in chronic renal failure suppresses PTH release by increasing the production of $1,25(OH)_2D_3$ [31, 32]. To gain further insight into the mechanism by which low phosphate diet prevents secondary hyperparathyroidism, studies were performed [33] in 7 normal dogs before and after the induction of uremia. The dogs were fed a low phosphate diet for one month. In the first month of low phosphate diet, ionized calcium did not change significantly. In spite of renal insufficiency, amino terminal iPTH became undetectable in 94% of 148 samples. Of interest, serum levels of $1,25(OH)_2D_3$ decreased from a mean of 21.1 ± 1.8 pg/ml (normal GFR) to 8.8 ± 0.9 pg/ml (1 month of renal failure). The suppression of PTH was not due to an increase in the levels of $1,25(OH)_2D_3$. Thus, a low phosphate diet prevented the development of secondary hyperparathyroidism with renal insufficiency. In addition to the well-known effect of phosphate in the regulation of $1,25(OH)_2D_3$, a low phosphate diet may have an effect on the secretion of PTH on a long-term basis. Potentially, phosphate may affect phospholipid composition of the parathyroid cell membrane, cytosolic calcium and calcium fluxes in the parathyroid cells and/or regulation of $1,25(OH)_2D_3$ receptors in the parathyroid glands, thus, indirectly phosphate may regulate the secretion of PTH. Further studies are necessary to precisely characterize the mechanism by which phosphate contributes to the regulation of PTH secretion in chronic renal failure.

References

1 Keutmann, H.T.; Sauer, M.M.; Hendry, G.N.; Potts, J.T., Jr.: Complete amino-acid sequence of human parathyroid hormone. Biochemistry *17:* 5723–5729 (1978).

2 Habener, J.F.; Potts, J.T., Jr.; Rich, A.: Pre-proparathyroid hormone: evidence for an early biosynthetic precursor of proparathyroid hormone. J. biol. Chem. *251:* 3893–3989 (1976).

3 Habener, J.F.; Rosenblatt, M.; Kemper, B.; Potts, J.T., Jr.: Pre-proparathyroid hormone: amino acid sequence, chemical synthesis, and some biological studies of the precursor region. Proc. natn. Acad. Sci. USA *75:* 2616–2620 (1978).

4 Chu, L.L.; MacGregor, R.R.; Hamilton, J.W.; Cohn, D.V.: Conversion of proparathyroid hormone to parathyroid hormone: the use of amines as specific inhibitors. Endocrinology *95:* 1431–1438 (1974).

5 Sherwood, L.M.; Mayer, G.P.; Ramberg, C.F.: Regulation of parathyroid hormone

secretion: proportional control by calcium, lack of effect of phosphate. Endocrinology *83:* 1043–1051 (1968).

6 Sherwood, L.M.; Herrman, I.; Barrett, C.A.: Parathyroid hormone secretion in vitro: regulation by calcium and magnesium ions. Nature *225:* 1056–1057 (1970).

7 Brown, E.M.; Gardner, D.G.; Windeck, R.A.; Auerbach, G.D.: Relationship of intracellular 3′,5′-adenosine monophosphate accumulation to parathyroid hormone release from dispersed bovine parathyroid cells. Endocrinology *103:* 2323–2333 (1978).

8 Morrissey, J.J.; Cohn, D.V.: Regulation of secretion of parathormone and secretory protein-I from separate intracellular pools by calcium, dibutyryl cyclic AMP, and (1)-isoproterenol. J. Cell Biol. *82:* 93–102 (1979).

9 Dufresne, L.R.; Gitelman, H.J.: A possible role of adenyl cyclase in the regulation of parathyroid activity by calcium; in Talmadge, Munson, Calcium, parathyroid hormone and the calcitonins, pp. 202–206 (Excerpta Medica, Amsterdam 1972).

10 Rodriguez, J.H.; Morrison, A.; Slatopolsky, E.; Klahr, S.: Adenyl cyclase of human parathyroid glands. J. clin. Endocr. Metab. *47:* 319–325 (1978).

11 Brown, E.M.; Gardner, D.G.; Aurbach, G.D.: Effects of the calcium ionophore A23187 on dispersed bovine parathyroid cells. Endocrinology *106:* 133–139 (1980).

12 Brown, E.M.: Relationship of 3′,5′-adenosine monophosphate accumulation to parathyroid hormone release in dispersed cells from pathological human parathyroid tissue. J. clin. Endocr. Metab. *52:* 961–968 (1981).

13 Schatzmann, H.J.: Active calcium transport and Ca^{++} activated ATPase in human red cells; in Bronner, Kleinzeller, Current topics in membranes and transport, vol. 6, pp. 125–168 (Academic Press, New York 1975).

14 Blaustein, M.; Ratzlaff, R.W.; Kendrick, N.K.: The regulation of intracellular calcium in presynaptic nerve terminals. Ann. N.Y. Acad. Sci *307:* 195–202 (1978).

15 Taylor, A.; Windhager, E.E.: Possible role of cytosolic calcium and Na-Ca exchange in regulation of transepithelial sodium transport. Am. J. Physiol. *236:* F505–511 (1979).

16 Rothstein, M.; Morrissey, J.; Slatopolsky, E.; Klahr, S.: The role of Na^{+}-Ca^{++} exchange in parathyroid hormone secretion. Endocrinology *111:* 225–230 (1982).

17 Brown, E.M.; Brennan, M.F.; Hurwitz, S.: Dispersed cells prepared from human parathyroid glands: distinct calcium sensitivity of adenomas vs. primary hyperplasia. J. clin. Endocr. Metab. *46:* 267–275 (1978).

18 Brown, E.M.; Wilson, R.E.; Eastman, R.C.; Pallota, J.; Marynick, S.P.: Abnormal regulation of parathyroid hormone release by calcium in secondary hyperparathyroidism due to chronic renal failure. J. clin. Endocr. Metab. *54:* 172–179 (1982).

19 Oldham, S.B.; Fischer, J.A.; Shen, L.H.; Arnaud, C.D.: Isolation and properties of a calcium-binding protein from porcine parathyroid glands. Biochemistry *13:* 4790–4796 (1974).

20 Brumbaugh, P.F.; Hughes, M.R.; Haussler, M.R.: Cytoplasmic and nuclear binding components for 1a,25-dihydroxyvitamin D_3 in chick parathyroid glands. Proc. natn. Acad. Sci. USA *72:* 4871–4875 (1975).

21 Chertow, B.S.; Baylink, D.J.; Wergedal, J.E.; Su, M.H.H.; Norman, A.W.: Decrease in serum immunoreactive parathyroid hormone in rats and in parathyroid hormone

secretion in vivo by 1,25-dihydroxycholecalciferol. J. clin. Invest. *56:* 668–678 (1975).
22 Oldham, S.B.; Smith, R.; Hartenbower, D.L.; Henry, H.L.; Norman, A.W.; Coburn, J.W.: The acute effects of 1,25-dihydroxycholecalciferol on serum immunoreactive parathyroid hormone in the dog. Endocrinology *104:* 248–254 (1979).
23 Brumbaugh, P.F.; Haussler, D.H.; Bressler, R.; Haussler, M.R.: Radio-receptor assay for 1,25-hydroxyvitamin D_3. Science *183:* 1089–1091 (1974).
24 Slatopolsky, E.; Gray, R.; Adams, N.; Lewis, J.; Hruska, K.; Martin, K.; Klahr, S.; DeLuca, H.; Lemann, J.: The pathogenesis of secondary hyperparathyroidism in early renal failure; in Norman, Vitamin D: basic research and its clinical application (de Gruyter, Berlin 1979).
25 Madsen, S.; Olgaard, K.; Ladefoged, J.: Suppressive effect of 1,25-dihydroxyvitamin D_3 on circulating parathyroid hormone in acute renal failure. J. clin. Endocr. Metab. *53:* 823–827 (1981).
26 Slatopolsky, E.; Weerts, C.; Thielan, J.; Horst, R.; Harter, H.; Martin, K.: Marked suppression of secondary hyperparathyroidism by intravenous administration of 1,25-dihydroxycholecalciferol in uremic patients. J. clin. Invest. *74:* 2136–2143 (1984).
27 Chan, Y.L.; McKay, C.; Dye, E.; Slatopolsky, E.: The effect of 1,25-dihydroxycholecalciferol on parathyroid hormone secretion by monolayer cultures of bovine parathyroid cells. Calcif. Tissue Int. *38:* 27–32 (1986).
28 Silver, J.; Russell, J.; Sherwood, L.M.: Regulation by vitamin D metabolites of messenger ribonucleic acid for preproparathyroid hormone in isolated bovine parathyroid cells. Proc. natn. Acad. Sci. USA *82:* 4270–4273 (1985).
29 Silver, J.; Naveh-Many, T.; Mayer, H.; Schmelzer, H.J.; Popovtzer, M.M.: Regulation by vitamin D metabolites of parathyroid hormone gene transcription in vivo in the rat. J. clin. Invest. *78:* 1296–1301 (1986).
30 Lopez-Hilker, S.; Galceran, T.; Chan, Y.L.; Rapp, N.; Martin, K.; Slatopolsky, E.: Hypocalcemia may not be essential for the development of secondary hyperparathyroidism in chronic renal failure. J. clin. Invest. *78:* 1097–1102 (1986).
31 Portale, A.A.; Booth, B.E.; Halloran, B.P.; Morris, C., Jr.: Effect of dietary phosphorus on circulating concentrations of 1,25-dihydroxyvitamin D and immunoreactive parathyroid hormone in children with moderate renal insufficiency. J. clin. Invest. *73:* 1580–1589 (1984).
32 Llach, F.; Massry, S.G.: On the mechanism of secondary hyperparathyroidism in moderate renal insufficiency. J. clin. Endocr. Metab. *61:* 601–606 (1985).
33 Lopez-Hilker, S.; Rapp, N.; Martin, K.; Slatopolsky, E.: On the mechanism of the prevention of secondary hyperparathyroidism by phosphate restriction (Abstract). Kidney int. *29:* 164a (1986).

Eduardo Slatopolsky, MD, Renal Division, Department of Medicine,
Washington University School of Medicine, St. Louis, MO 63110 (USA)

Contr. Nephrol., vol. 64, pp. 25–33 (Karger, Basel 1988)

Vitamin D and the Kidney: A Short Review

R. Bouillon

Laboratorium voor Experimentele Geneeskunde en Endocrinologie, Onderwijs en Navorsing, Leuven, Belgium

Vitamin D is an essential precursor substance in most vertebrates [1–4] and can either be provided by nutritional intake or synthesis in the skin under the influence of the shortest waves of the ultraviolet (UV) light. The nutritional intake of vitamin D varies widely from population to population and from individual to individual because vitamin D is only present in substantial amounts in some food (e.g. fatty fish, egg yolk, fatty dairy products or artificially vitamin-D-enriched food), used in variable amounts by most people. Nutritional vitamin D is absorbed by the gut, then transported by the chylomicrons to the blood stream (fig. 1) and thereafter either stored in several tissues or hydroxylated in the liver into 25-hydroxyvitamin D (25-OHD) [5]. The synthesis of vitamin D in the skin requires the activation of 7-dehydrocholesterol, present in the Malpighian layers of the skin, by UV light with a short wavelength (±280–310 nm) whereby the end product, previtamin D, is either slowly thermally isomerized into vitamin D or further irradiated into metabolically inactive compounds [6].

Vitamin D is removed from the skin by binding to a specific vitamin-D-binding serum protein, known as group-specific component or DBP [4] which has an α-globulin mobility (in man) and a molecular weight somewhat lower than that of albumin to which it is genetically related [7]. The liver converts vitamin D into 25-OHD by a microsomal hydroxylase system but in cases of excess vitamin D a mitochondrial 25-hydroxylation may also become active. Although 1,25-$(OH)_2D$ may exert a negative feedback inhibition on the hepatic production of 25-OHD, the serum 25-OHD levels usually reflect the nutritional access to vitamin D [8]. Serum 25-OHD is bound with high affinity to the high capacity serum

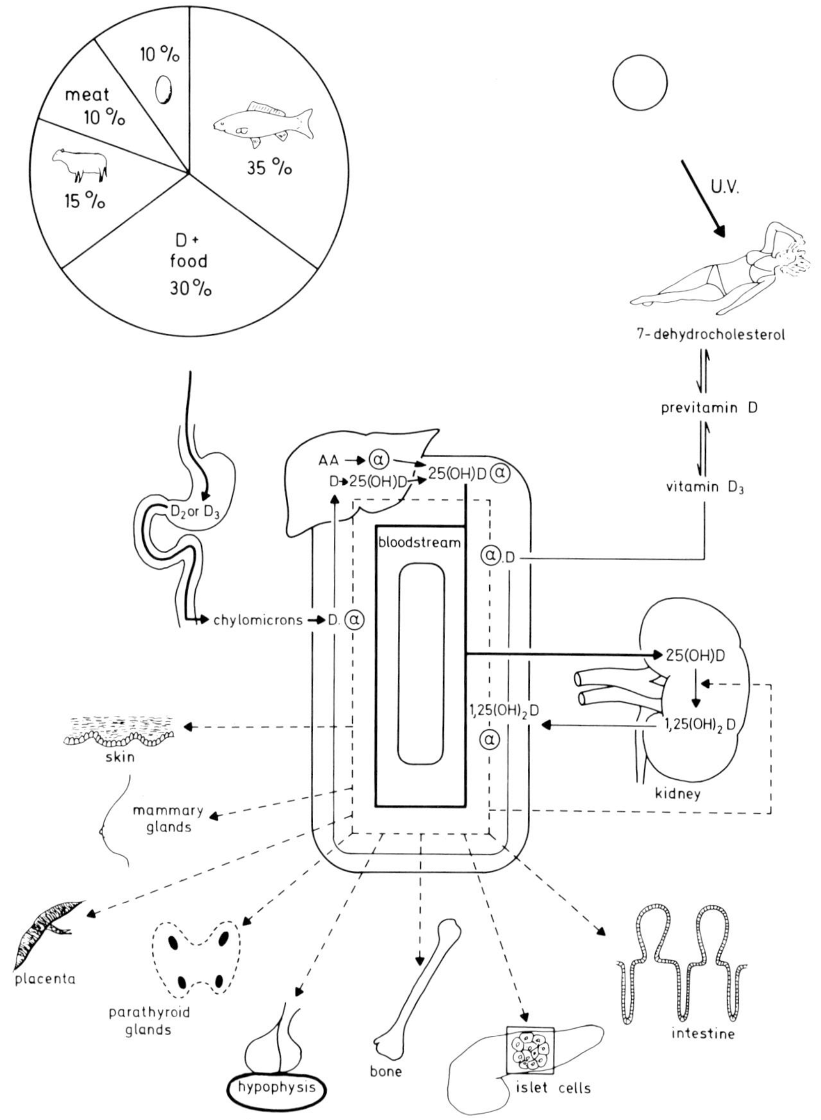

Fig. 1. Overview of the metabolism, transport and actions of vitamin D.

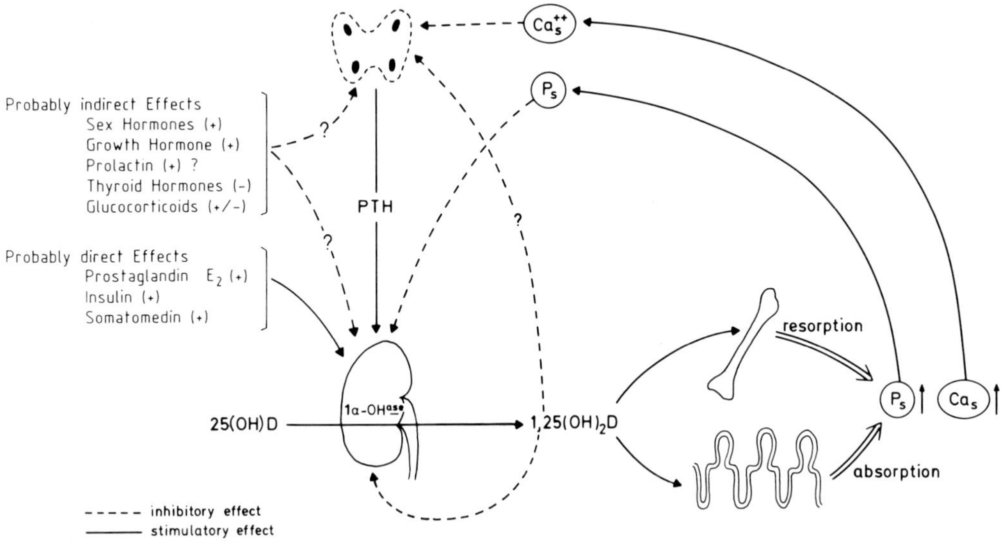

Fig. 2. Summary of the feed-back regulation of the renal 25-hydroxyvitamin D-1α-hydroxylase activity.

DBP pool so that the half-life of 25-OHD exceeds several weeks. About 1 μg of 25-OHD is daily transformed to 1,25-$(OH)_2D$ by a specific mitochondrial 1,25-OHD-1α-hydroxylase system present in the proximal tubular cells of the kidney [9]. This kidney enzyme is stimulated by parathyroid hormone (PTH), low serum (and intracellular?) phosphate concentration and is suppressed by its end product [1,25-$(OH)_2D$] and maybe also by high serum calcium. Other humoral or hormonal substances may also influence the renal 1α-hydroxylase activity but their direct effects are more disputed (fig. 2). In any way, the 1,25-$(OH)_2D$ production is tightly feedback-regulated by a long feedback system using serum phosphate, calcium and PTH secretion, an intermediate system using a direct inhibition of PTH synthesis and a short feedback system involving the suppression of its own synthesis (inhibition of the 1α-hydroxylase enzyme synthesis) and enhancing its own degradation (switching on the synthesis of the 24-hydroxylase enzyme).

Serum 1,25-$(OH)_2D$ is also bound to serum DPB but with a much lower affinity so that it can more easily reach and bind to the high affinity,

Table I. Summary of known and possible target tissues for vitamin D, containing high specificity receptors for 1,25-dihydroxyvitamin D

1 Tissues with known calcium transport	2 Endocrine tissues
Intestinal mucosa	Endocrine pancreas
Bone	Pituitary gland
Placenta	Parathyroid gland
Chorioallantois membrane	Testis/ovary
Mammary gland	
3 Immune cells	4 Others
Thymus	Muscle
Monocytes/macrophages	Malignant tumors
Activated B and T cells	(e.g. breast carcinoma)

low capacity nuclear receptors with high specificity for 1,25-$(OH)_2D$. These receptors are present in a large number of tissues, either specialized in calcium transport, hormone synthesis in cells using intracellular calcium as second messenger or other cells where calcium may have an important functional role (table I).

Although there are some arguments in favor of a direct and immediate effect of vitamin D on cell membrane function or fluidity [10], the most convincing mode of action of physiologic amounts of 1,25-$(OH)_2D$ consists of nuclear binding and activation of genomic functions, similar to the action of other steroid or thyroid hormones. Vitamin D itself and 25-OHD would therefore only function as hormonal precursor to avoid sudden deficiency or excess similar to iodine and thyroid hormone metabolism (fig. 3).

The best known effect of vitamin D is its role in the calcium and bone homeostasis by enhancing the intestinal absorption of calcium and phosphate and by its direct or indirect effect on bone resportion and bone formation. There is now also increasing evidence for a role of vitamin D in the hormonal secretion of insulin [11], PTH, prolactin and maybe other hormones. Moreover, an increasing list of in vitro and in vivo effects of vitamin D on the immune cells has come to light [12]. The active vitamin D hormone could thus be more than just a calcium-regulating substance and have a growth and differentiation effect in many cells.

How is the vitamin D metabolism and action affected by kidney diseases? The supply of vitamin D is usually not severely impaired in chronic

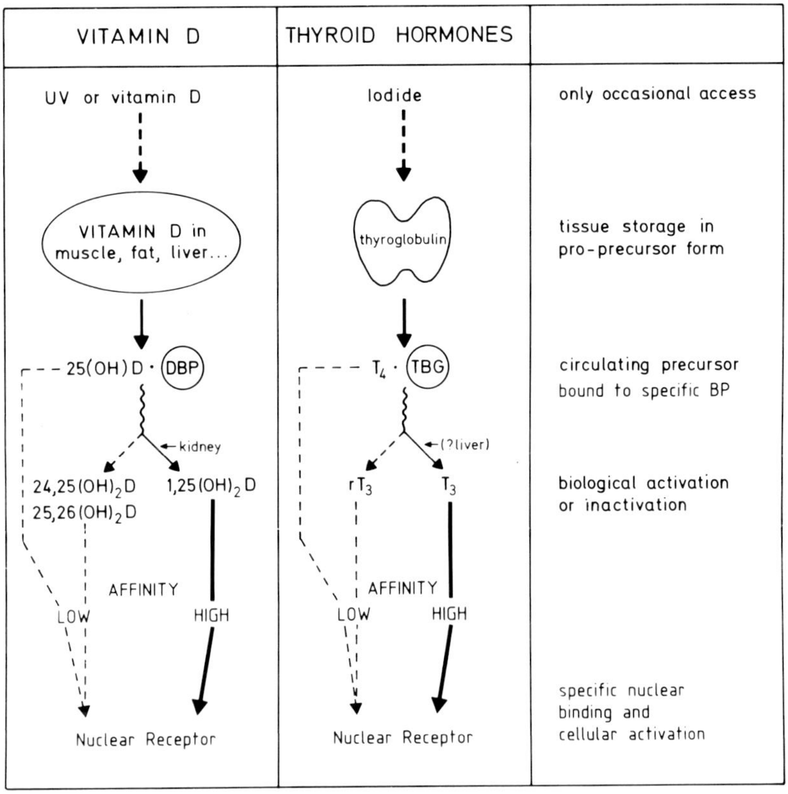

Fig. 3. Pathways of storage, activation and degradation of vitamin D, compared to the metabolism of iodine and the thyroid hormones.

renal failure. Indeed, the intestinal absorption of vitamin D is not altered but the skin synthesis of vitamin D may be less efficient due to increased skin pigmentation [13]. Occasional vitamin D deficiency in patients with chronic renal failure has therefore only been observed in countries with limited sun exposure or in patients with severely reduced mobility. On the other hand, the already long half-life of serum 25-OHD is increased in chronic renal failure, probably by decreased metabolic clearance in the kidney, so that this phenomenon may compensate the decreased skin production (table II). Only in cases of severe proteinuria (nephrotic syndrome) a true deficiency of vitamin D and 25-OHD will be found due to

Table II. Vitamin D metabolism in chronic renal failure

1	Vitamin D Normal intestinal absorption Decreased skin synthesis
2	25-Hydroxyvitamin D Normal liver 25-hydroxylase activity Decreased catabolism with doubling of the t½ Normal (USA) or decreased (UK) serum concentrations More severe deficiency in case of nephrotic syndrome
3	1,25-Dihydroxyvitamin D Decreased serum concentration if GFR falls below 50 ml/min due to structural (tubular mass) and functional (high phosphate concentration, acidosis and aluminum) abnormalities Not compensated by extrarenal 1,25-$(OH)_2D$ production
4	Other vitamin D metabolites Decreased 24,25-$(OH)_2D$ concentrations (mainly kidney-produced) Unchanged 25,26-$(OH)_2D$ concentrations
5	Consequences Decreased intestinal calcium absorption Secondary hyperparathyroidism Bone mineralization defects

the urinary loss of DBP together with its ligands. This loss may exceed the supply of vitamin D so that true deficiency of protein-bound and free 1,25-$(OH)_2D$ occurs occasionally resulting in osteomalacia [14, 15].

The production of 1,25-$(OH)_2D$ will decrease since the total tubular mass of 1α-hydroxylase activity decreases as the kidney disease progresses [16–20]. This may initially be partially compensated by secondary hyperparathyroidism, which itself is explained by hypocalcemia and by loss of direct inhibition of PTH synthesis [21]. If glomerular filtration rate (GFR) falls below 50–60 ml/min, the decreasing serum 1,25-$(OH)_2D$ will result in decreasing intestinal calcium absorption, further aggravating the secondary hyperparathyroidism and deteriorating the bone abnormalities due to this secondary hyperparathyroidism. It is unclear whether the decrease in 1,25-$(OH)_2D$ or 24,25-$(OH)_2D$ itself is a significant contributing factor to the renal osteodystrophy, since deficiency of these vitamin D hormones would have to result in osteomalacia which is only rarely observed in the absence of aluminum toxicity or true vitamin D and 25-OHD deficiency.

Although initially the kidney was described as the unique origin of 1,25-$(OH)_2$D [9], it is now clear that other tissues are also capable of producing this hormonal metabolite. Except for the placenta, bone, lymphocytes or activated monocytes-macrophages and more questionably cartilage and intestine have been reported to be able, at least in vitro, to transform 25-OHD into more active metabolites. This contrasts with the repeated observations that in vivo no transformation of radioactive vitamin D into 1,25-$(OH)_2$D can be observed after bilateral nephrectomy [9, 21]. When, however, 1,25-$(OH)_2$D is measured in the serum of anephric humans, small but detectable amounts of 1,25-$(OH)_2$D have been observed. These observations were initially considered as possible artifacts of the assay methodology but more recent studies undoubtedly demonstrated this phenomenon [23, 24]. Such small concentrations of 1,25-$(OH)_2$D can, however, only be found if the substrate concentration (25-OHD) is increased into the high-normal or toxic range, similar to what has been observed in primary hyperparathyroidism. These apparent contradictions between results obtained in anephric animals and human patients with chronic renal failure can probably best be explained by the difference in duration of observations as all animal studies have been short term whereas the observations in man showed that several days were required before serum 1,25-$(OH)_2$D increased after 25-OHD loading [24].

It is also apparent from such observations that the kidney is normally the only source of circulating 1,25-$(OH)_2$D, thereby not excluding that bone or other cells may produce sufficient 1,25-$(OH)_2$D for local action, explaining why true osteomalacia is rare in the absence of aluminium toxicity.

The disturbed vitamin D metabolism in chronic renal failure or nephrotic syndrome requires some specific therapeutic measures:

(1) In the case of chronic nephrotic syndrome, the serum 25-OHD should be normalized by sufficient sun exposure or oral intake of vitamin D or 25-OHD to compensate the urinary loss of vitamin D metabolites.

(2) In the case of chronic renal failure, 25-OHD deficiency should be monitored and corrected in countries with low sun exposure or in the case of suspicion of true vitamin D deficiency. Secondary hyperparathyroidism is the main factor responsible for renal osteodystrophy in the absence of aluminum toxicity and can be avoided by several (combined)

techniques including a decreased intake or intestinal absorption of phosphate, an increase of calcium intake and absorption, and by increasing the serum 1,25-$(OH)_2D$ concentrations. Although this could be achieved in some patients by increasing their serum 25-OHD and exploiting the remaining renal and mainly extrarenal 1α-hydroxylase activity, it nevertheless seems more logical to use a physiologic replacement therapy with 1α-hydroxylated vitamin D metabolites [25–27].

References

1 Brommage, R.; DeLuca, H.F.: Evidence that 1,25-dihydroxyvitamin D_3 is the physiologically active metabolite of vitamin D_3. Endocr. Rev. *6:* 491–511 (1985).

2 Norman, A.W.; Roth, J.; Orci, L.: The vitamin D endocrine system: steroid metabolism, hormone receptors and biological response (calcium-binding proteins). Endocr. Rev. *3:* 331–336 (1982).

3 Fraser, D.R.: Regulation of the metabolism of vitamin D. Physiol. Rev. *60:* 551 (1980).

4 Bouillon, R.; Van Baelen, H.: The transport of vitamin D: significance of free and total concentrations of vitamin D metabolites; in Norman et al., Vitamin D, chemical, biochemical and clinical endocrinology of calcium metabolism, pp. 1181–1186 (de Gruyter, Berlin 1982).

5 Dueland, S.; Bouillon, R.; Van Baelen, H.; Pedersen, J.I.; Helgerud, P.; Drevon, C.A.: Binding protein for vitamin D and its metabolites in rat mesenteric lymph. Am. J. Phys.: E1–E5 (1985).

6 Holick, M.F.: Vitamin D photobiology: recent advances in the biochemistry and some clinical applications; in Norman et al., Vitamin D, chemical, biochemical and clinical update, pp. 219–228 (de Gruyter, Berlin 1985).

7 Cooke, N.E.: Serum vitamin D binding protein is a third member of the albumin and α-fetoprotein gene family. J. clin. Invest. *76:* 2420–2424 (1975).

8 Bouillon, R.: Radiochemical assays for vitamin D metabolites: technical possibilities and clinical applications. J. Steroid Biochem. *19:* 921–927 (1983).

9 Fraser, D.R.; Kodicek, E.: Unique biosynthesis by kidney of a biologically active vitamin D metabolite. Nature *228:* 764–766 (1970).

10 Yoshimoto, Y.; Norman, A.W.: Biological activity of vitamin D metabolites and analogs: dose-response study of ^{45}Ca transport in an isolated chick duodenum perfusion system. J. Steroid Biochem. *25:* 905–909 (1986).

11 Nyomba, G.L.; Bouillon, R.; De Moor, P.: Influence of vitamin D status on insulin secretion and glucose tolerance in the rabbit. Endocrinology *115:* 191–197 (1984).

12 Provvedini, D.M.; Tsoukas, C.D.; Deftos, L.J.; Manolagas, S.C.: 1,25-Dihydroxyvitamin D_3 receptors in human leukocytes. Science *221:* 1181 (1983).

13 Hollis, B.W.; Jacob, A.I.; Sallman, A.; Sanitz, Z.; Lambert, P.W.: Circulating vitamin D and its photoproduction in uremia; in Norman et al., Vitamin D chemical, biochemical and clinical endocrinology of calcium metabolism, pp. 1157–1160 (de Gruyter, Berlin 1982).

14 Schmidt-Gayk, H.; Schmitt, W.; Grawunder, C.; Ritz, E.; Tschöpe, W.; Pietsch, V.; Andrassay, K.; Bouillon, R.: 25-Hydroxy-vitamin D in nephrotic syndrome. Lancet *ii:* 105–108 (1977).

15 Auwerx, J.; De Keyser, L.; Bouillon, R.; De Moor, P.: Decreased free 1,25-dihydroxycholecalciferol index in patients with the nephrotic syndrome. Nephron *42:* 231–235 (1986).

16 Fournier, A.; Boudailliez, B.; Tolani, M.; Ulmann, A.; Grégoire, I.; Marié, A.; Morinière, P.; Sebert, J.L.: Vitamine D et ostéodystrophie rénale; in Fournier et al., Vitamine D et maladies des os et du métabolisme minéral, pp. 170–246 (Masson, Paris 1984).

17 Howard, M.M.; Cushner, M.; Adams, N.D.: Renal osteodystrophy. Pathogenesis and treatment. Am. J. med. Sci. *290:* 234–245 (1985).

18 Chesney, R.W.; Hamstra, A.J.; Mazess, R.B.; Rose, P.; DeLuca, H.F.: Circulating vitamin D metabolite concentrations in childhood renal diseases. Kidney int. *21:* 65–69 (1982).

19 Coburn, J.W.; Norris, K.C.; Salusky, I.B.; Andress, D.L.; Crooks, P.W.; Ott, S.M.; Hercz, G.; Nebeker, H.G.; Milliner, D.A.; DiDominico, N.C.; Sherrard, D.J.: Renal osteodystrophy: views on pathogenesis and management – 1985; in Norman et al., Vitamin D, chemical, biochemical and clinical update, pp. 926–934 (de Gruyter, Berlin 1985).

20 Portale, A.A.; Booth, B.E.; Tsai, H.C.; Morris, R.C.: Reduced plasma concentration of 1,25-dihydroxyvitamin D in children with moderate renal insufficiency. Kidney int. *21:* 627–632 (1982).

21 Slatopolsky, E.; Weerts, C.; Thielan, J.; Horst, R.; Harter, H.; Martin, K.J.: Marked suppression of secondary hyperparathyroidism by intravenous administration of 1,25-dihydroxycholecalciferol in uremic patients. J. clin. Invest. *74:* 2136–2143 (1984).

22 Reeve, L.; Tanaka, Y.; DeLuca, H.F.: Studies on the site of 1,25-dihydroxyvitamin D_3 synthesis in vivo. J. biol. Chem. *258:* 3615–3617 (1983).

23 Jongen, M.J.M.; Vijgh, W.J.F. van der; Lips, P.; Netelenbos, J.C.: Measurement of vitamin D metabolites in anephric subjects. Nephron *36:* 230–234 (1984).

24 Fraher, L.J.; Adami, S.; Papapoulos, S.E.; Sudan, H.L.; McGongile, R.J.; Parsons, V.; O'Riordan, J.L.H.: Evidence for extrarenal metabolism of 25-hydroxyvitamin D_3 in man. Clin. Sci. *71:* 89–95 (1986).

25 Chan, J.C.M.; Oldham, S.B.; Holick, M.F.; DeLuca, H.F.: 1-α-Hydroxyvitamin D_3 in chronic renal failure. J. Am. med. Ass. *234:* 47–52 (1975).

26 Robitaille, P.; Marie, P.J.; Delvin, E.E.; Lortie, L.; Glorieux, F.H.: Renal osteodystrophy in children treated with 1,25-dihydroxycholecalciferol (1,25-$[OH]_2D_3$). Acta paediat. scand. *73:* 315–324 (1984).

27 Cundy, T.; Kanis, J.A.; Earnshaw, M.; Woods, C.G.: Comparative effects of alfacalcidol and parathyroidectomy with vitamin D in hyperparathyroid renal bone disease. Q. Jl Med. *231:* 659–670 (1986).

R. Bouillon, MD, Laboratorium voor Experimentele Geneeskunde en Endocrinologie, Onderwijs en Navorsing, Gasthuisberg, B-3000 Leuven (Belgium)

Contr. Nephrol., vol. 64, pp. 34–42 (Karger, Basel 1988)

Disturbed Growth in Uremia: Are Hormonal Factors Responsible?

Otto Mehls [a], *Eberhard Ritz* [b], *Jürgen Merke* [b], *Udo Heinrich* [a], *Günter Klaus* [a]

[a]Department of Pediatrics and [b]Internal Medicine, University of Heidelberg, Heidelberg, FRG

The failure of longitudinal growth is a frequent finding in renal failure. Its pathogenesis is complex. Nutritional, hormonal and other factors have been recognized to play a role [1]. In this overview, we review some recent findings relating to the roles of calcium regulating hormones and the growth hormone (GH)/somatomedin system in the growth disturbances of uremia.

Calcium Regulating Hormones and Growth Failure of Uremia

The action of parathyroid hormone (PTH) on growth has not been completely clarified. It is commonly assumed that children with hypoparathyroidism grow poorly. However, there are no published data in the literature which document the type of growth disturbance encountered in children with hyperparathyroidism. Specifically, the separate roles of hypercalcemia and PTH excess have not been distinguished to date.

Two clinical observations argue for a permissive or even stimulatory role of PTH on growth. When hypoparathyroid children are rendered normocalcemic with vitamin D therapy, they fail to grow normally, possibly because PTH is lacking [2]. A mitogenic action of PTH on chondrocytes would be consistent with previous observations of McManus [3] that PTH triggers mitoses in various cells, e.g. thymocytes, osteogenic cells, hematopoetic cells [3] and promotes compensatory renal growth [4].

On the other hand, children with pseudohypoparathyroidism and osteitis fibrosa are usually tall [5]. Such children have high circulating

Table I. Growth cartilage cAMP (pmol/mg protein)

	Control	Uremia
In vivo (n=10 per group)		
Basal	2.0±0.1	1.8±0.1
20 IU PTH	9.1±1.4	6.6±0.6
40 IU PTH	23.9±3.1	11.4±1.0
In vitro (n=9 per group)		
Basal	20.0±0.4	18.8±0.9
0.1 IU PTH/ml	108.3±10.8	61.1±8.8
1.0 IU PTH/ml	147.2±13.8	77.7±5.5

PTH levels. Since their skeleton (and possibly the growth apparatus) is responsive to PTH despite PTH unresponsiveness of the kidneys, high growth rates and tall stature may be caused by elevated PTH levels.

The action of PTH on the growth apparatus in uremia has not been well characterized. In a previous study [6] we found a diminished response of growth cartilage adenosine 3′:5′-cyclic phosphate (cAMP) to administration of PTH in vivo or addition of PTH in vitro. Some of these results are summarized in table I.

In the light of the above considerations, it is not surprising that we failed to show a correlation between the severity of X-ray evidence of osteitis fibrosa of uremic children on the one hand and growth rate on the other. Similarly, Hodson et al. [7] also failed to see a relation between osteitis fibrosa in bone biopsies and growth rate.

Stunting of growth is a hallmark of vitamin-D-dependent rickets type I [8]. Accumulation of labelled $1,25(OH)_2D_3$ in the growth apparatus [9], although not consistent in all studies [10] and metabolism of $1,25(OH)_2D_3$ in chondrocytes of epiphyseal cartilage [11] have been demonstrated. Using the $1,25(OH)_2D_3$ receptor assay [12] in preliminary studies, we have been able to demonstrate receptors for $1,25(OH)_2D_3$ in growth cartilage cells of normal animals (sensitivity approximately 3 fmol specific binding of ^3H $1,25[OH]_2D_3$/mg protein), which have also been described for bone cells [13]. Therefore, it is very likely that chondrocytes have the molecular machinery to recognize, and to respond to, circulating $1,25(OH)_2D_3$.

Since bioavailability of the active vitamin D metabolite $1,25(OH)_2D_3$ is diminished in uremia, it was assumed that deficiency of $1,25(OH)_2D_3$ was an important cause of impaired growth, as it is in vitamin-D-dependent rickets type I. This hypothesis was supported by the observation of Chesney et al. [14] that $1,25(OH)_2D_3$ promoted growth in some uremic children who had failed to respond to high doses of vitamin D_3. This observation was also supported by similar reports of other authors [15]. In most of these studies it is impossible to separate the action of $1,25(OH)_2D_3$ from a fortuitous pubertal growth spurt. None of the above studies found normalization of growth by $1,25(OH)_2D_3$ and the growth-promoting effect of $1,25(OH)_2D_3$ in the original communication of Chesney et al. [14] was only transitory, since follow-up of his children showed that growth curves continued to parallel the third percentile of age-corrected growth charts [16]. Indeed, several well controlled prospective long-term studies [17] failed to show a consistent effect of $1,25(OH)_2D_3$ or 1α-hydroxycalciferol on growth and this is also in agreement with our own experience.

In experimental studies, we were able to show that the growth deficit of uremic animals [18] could be overcome in part by administration of $1,25(OH)_2D_3$ or doses of cholecalciferol equipotent with respect to their calcemic action [19, 20]. This is illustrated by figure 1 taken from our recent review [20].

On the whole, the effects of calcium-regulating hormones on growth failure in uremia are controversial and are likely to be modest in magnitude. It must be admitted, however, that advanced osteitis fibrosa by causing destructive lesions in the metaphyseal growth zone may cause cessation of growth secondary to destruction of the architecture of the growth apparatus [21].

The Growth Hormone/Somatomedin System

The action of GH on growth zone has not been fully elucidated. It is commonly assumed that the effect of GH on growth is mediated by a GH-dependent effector polypeptide, somatomedin C/insulin-like growth factor I (IGF I), which is synthesized in the liver and released into the circulation. Several recent findings suggest an alternative, or complementary, mechanism, i.e. local action of GH on growth cartilage [22] or interaction of locally synthesized GH-dependent IGF with GH on growth car-

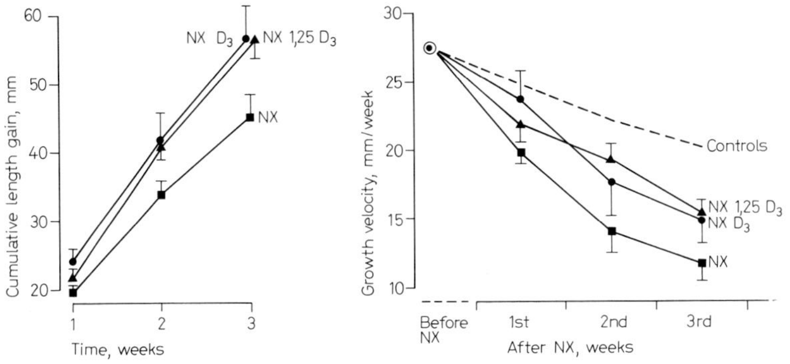

Fig. 1. In uremic rats growth is increased by vitamin D_3 (NX D_3) and by 1,25-vitamin D_3 (NX 1,25D_3). The effect of 1,25D_3 is not superior to the effect of D_3. Both drugs improve growth velocity but do not completly correct (right pannel). ±SEM.

tilage [23, 24]. Such interaction has led to the proposal of a 'dual effector theory' for GH response in various tissues including epiphyseal growth plates [25]. This notion implies that GH stimulates differentiation of chondrocyte precursor cells (prechondrocytes) in the growth plate causing activation of genes for growth factors like IGF I during the process of cell differentiation.

In uremic animals [26] and uremic children [27] or adults [28], fasting GH levels are elevated and a paradoxical rise of GH was noted after intravenous administration of glucose. Secretory stimuli like *L*-dopa, insulin-induced hypoglycemia or thyrotropin-releasing hormone (TRH) caused an exaggerated secretion of GH with delayed but increased peak secretion and increased area under the curve. These observations suggest an abnormal balance between hypothalamic stimulatory and inhibiting factors for GH secretory control [29]. The kidney is a catabolic site for GH [30]. However, in uremic patients somatostatin causes a rapid decrease of circulating GH levels [31]. This observation suggests that elevated fasting GH levels are primarily due to increased GH secretion.

The issue is much more confused with respect to somatomedin/IGF levels in uremia. Early studies using somatomedin bioassays were subject to artefacts. Low incorporation of radiosulfate into test cartilage was later recognized to be due to increased sulfate levels in serum of uremic children. However, more recently, bioassayable somatomedin activity was noted even when corrections for sulfate concentrations were used or when ^{3}H-radiothymidine incorporation was measured. Such inhibition was assumed to be due to the accumulation of low molecular weight circulating somatomedin inhibitors in uremia [32]. However, the factors have not been further characterized and the relevance of such in vitro observations to in vivo action of somatomedin remains to be clarified.

Circulating IGF is bound to several species of binding proteins, in the rat, a 45000- and a 140,000-molecular weight binding protein. In humans with renal failure, an increase of both binding protein species has recently been described [33, 34]. Since somatomedin and its binding protein are cosynthesized in the liver in stoichiometric amounts, this observation may explain several reports of elevated IGF levels in uremia.

Two different somatomedin species circulate in blood, IGF I and IGF II. Their relative proportion is altered in uremia with a marked increase of IGF II and a variable increase of IGF I [34–37]. However, the methodology for IGF measurements is delicate and susceptible to artifacts, especially if acid chromatography techniques are not used.

Finally, diminished GH receptors have been described on hepatocyte membranes [35, 37] and diminished secretion from the liver of a GH-dependent acid protein has been described in experimental uremia [38]. These observations, together with high basal GH levels, are consistent with the notion of GH hyporesponsiveness in uremia.

Since recombinant human GH (hGH) has recently become available in relatively large amounts for use in humans, we examined whether growth failure in uremia could be reversed, at least in part, by hGH in doses which would be available for clinical use [39]. As shown in figure 2, subcutaneous administration of 1.4 U hGH per day by osmotic minipumps to subtotally nephrectomized (NX) uremic rats and their sham-operated pair-fed controls caused a continuous consistent increase of growth, i.e. of weight gain and length gain. This was associated with an increase in the food conversion ratio, i.e. the ratio of weight gain (g) per food intake (g) 0.146 ± 0.01 in NX animals with GH vs 0.065 ± 0.03 in NX animals with solvent; $p < 0.01$. Anabolism under the influence of supraphysiological doses of GH was further documented by increased muscle dry weight (m. soleus 268 ± 8 mg in NX animals with GH vs 247 ± 9 mg in NX animals plus solvent; $p < 0.05$) despite almost identical cumulative food intake. The stimulatory effect of supraphysiological doses of GH on growth and anabolism was observed although no increase occurred of circulating IGF I concentrations in either uremic animals

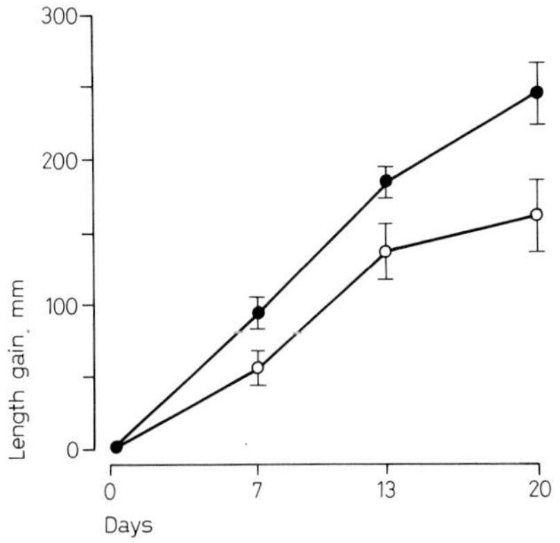

Fig. 2. In uremic rats growth is improved by recombinant GH (upper curve) administered i.p. at a dose of 1.4 IU/day by osmotic minipumps.

(174 ± 4 ng in NX + GH vs 188 ± 10 ng/ml in NX plus solvent) or sham-operated control animals (239 ± 14 vs 207 ± 6).

The data are consistent with one of the following hypotheses: (a) that high doses of GH overcome GH hyporesponsiveness of growth cartilage target cells, (b) that they induce local synthesis of GH-dependent mediators, e.g. IGF, or (c) that high GH concentrations restore responsiveness of cartilage cells to IGF.

Whichever explanations ultimately prove to be correct, the observations open the fascinating possibility to partially correct stunting of growth in uremic children by administration of pharmacological doses of recombinant GH. Clinical studies to test such an approach are currently underway.

References

1 Mehls, O.; Ritz, E.; Gilli, G.; Kreusser, W.: Growth in renal failure. Nephron *21:* 237 (1978).

2 Mehls, O.; Ritz, E.: Renal osteodystrophy; in Holliday, Barratt, Vernier, Ped. Nephrol. pp. 852–887 (Williams & Wilkins, Baltimore 1987).

3 Bingham, P.J.; Barzell, J.A.; Owen, M.: The effect of parathyroid extract on cellular activity and plasma calcium levels in vitro. J. Endocr. *45:* 387 (1969).

4 Bonjour, J.P.; Jobin, J.R.: Compensatory renal growth: modulation by calcium PTH and 1,25-$(OH)_2D_3$ (Personal communication). Kidney int. *29:* 1124–1130 (1986).

5 Parfitt, A.M.: Clinical and radiographic manifestations of renal osteodystrophy; in David, Calcium metabolism in renal failure and nephrolitiasis, pp. 145–196 (Wiley, New York 1977).

6 Kreusser, W.; Weinkauf, R.; Mehls, O.; Ritz, E.: Effect of parathyroid hormone, calcitonin and growth hormone on cAMP content of growth cartilage in experimental uremia. Eur. J. clin. Invest. *12:* 337–343 (1982).

7 Hodson, E.M.; Shaw, P.F.; Evans, R.A.; Dunstan, C.R.; Hills, E.E.; Wong, S.Y.P.; Rosenberg, A.R.; Roy, L.P.: Growth retardation and renal osteodystrophy in children with chronic renal failure. J. Pediat. *103:* 735 (1983).

8 Prader, A.; Illig, R.; Heierli, E.: Eine besondere Form der primären vitamin D-resistenten Rachitis mit Hypopocalciämie und autosomal dominantem Erbgang. Helv. paediat. Acta *16:* 452 (1961).

9 Dickson, I.; Maher, M.: The influence of vitamin D metabolites on collagen synthesis by chick cartilage in organ culture. J. Endocr. *105:* 79–85 (1985).

10 Narbaitz, R.; Stumpf, W.E.; Sar, M.; Huang, S.; De Luca, H.F.: Autoradiographic localization of target cells for 1,25-dihydroxyvitamin D_3 in bones from fetal rats. Calcif. Tissue Int. *35:* 177–182 (1983).

11 Garabedian, M.; Lieberherr, M.; Corvol, M.T.; Balsan, S.: Cellular and subcellular location of the 24,25$(OH)_2D_3$ formation in cultured cells from bone and cartilage; in Norman et al., Proc. 4th Workshop on Vitamin D, Berlin, pp. 391–397 (de Gruyter, Berlin 1979).

12 Merke, G.K.; Hügel, U.; Waldherr, R.; Ritz, E.: No 1,25-Dihydroxyvitamin D_3 receptors on osteoclasts of calcium-deficient chicken despite demonstrable receptors on circulating monocytes. J. clin. Invest. *77:* 312–314 (1986).

13 Chen, T.; Feldmann, D.: Regulation of 1,25-dihydroxyvitamin D_3 receptors in cultured mouse bone cells. J. biol. Chem. *256:* 5561–5566 (1981).

14 Chesney, R.W.; Moorthy, A.V.; Eismann, J.A.; Jax, D.K.; Mazzes, R.B.; De Luca, H.F.: Increased growth after long-term oral 1-alpha-25-vitamin D_3, in childhood renal osteodystrophy, New Engl. J. Med. *298:* 238 (1978).

15 Chan, J.C.M.; De Luca, H.F.: Growth velocity in a child on prolonged hemodialysis. Beneficial effect of 1-alpha-hydroxyvitamin D_3, J. Am. med. Ass. *238:* 2053 (1977).

16 Chesney, R.W.: 1,25-Dihydroxy-vitamin D_3 in the treatment of juvenile renal osteodystrophy; in Gruskin, Norman, Pediatric nephrology, p. 209 (Nijhoff, Den Haag 1981).

17 Bulla, M.; Delling, G.; Benz-Bohm, G.; Stock, G.J.; Sánchez de Reutter, A.; Ziegler, R.; Lühmann, H.; Severin, M.; Kalbitzer, E.; Manegold, C.: Renale Osteodystrophie bei Kindern. Therapieversuch mit 1,25-Dihydrosy-Cholecalciferol. Klin. Wschr. *58:* 511–519 (1980).

18 Mehls, O.; Ritz, E.; Gilli, G.; Bartholome, K.; Beisbarth, H.; Hohenegger, M.; Schafnitzel, W.: Nitrogen metabolism and growth in experimental uremia. Int. J. Pediatr. Nephrol. *1:* 34–41 (1980).

19 Mehls, O.; Ritz, E.; Gilli, G.; Wangdak, T.; Krempien, B.: Effect of vitamin D on growth in experimental uremia. Am. J. clin. Nutr. *31:* 1927–1931 (1978).

20 Mehls, O.; Ritz, E.: Skeletal growth in experimental uremia. Kidney int. *24:* suppl. 15, p. S53 (1983).
21 Krempien, B.; Mehls, O.; Ritz, E.: Morphological studies on pathogenesis of epiphyseal slipping in uremic children. Virchows Arch. Histol. Path. Anat. *362:* 129 (1974).
22 Isaksson, O.; Jansson, J.; Gause, J.: Growth hormone stimulates longitudinal bone growth directly. Science *216:* 1237–1239 (1982).
23 Slootweg, M.C.; Van Buul-Offers, S.C.; Duursma, S.A.: Mitogenic action of growth hormone on fetal chicken osteoblasts without production of somatomedins. 9th Int. Conf. on Calcium Regulating Hormones and Bone Metabolism, Nice 1986, p. 329.
24 Ernst, M.; Froesch, E.R.: Stimulatory effects of human growth hormone on osteoblastic cells (Ob) in a serum-free culture system. 9th Int. Conf. on Calcium Regulating Hormones and Bone Metabolism, Nice 1986, p. 311.
25 Green, H.; Morrikawa, M.; Nixon, T.: Actual effector theory of growth hormone action differentiation. Science *29:* 195–198 (1985).
26 Schalch, D.S.; Burstein, P.J.; Tewel, S.J.; Draznin, B.; Emler, D.A.: The effect of renal impairment on growth in the rat: relationship to malnutrition and serum somatomedin levels. Endocrinology *108:* 1683 (1981).
27 El-Bishti, M.; Counahan, R.; Bloom, S.; Chantler, C.: Hormonal and metabolic responses to intravenous glucose in children on regular hemodialysis. Am J. clin. Nutr. *31:* 1865–1869 (1978).
28 Samaan, N.; Fresman, R.: Growth hormone levels in severe renal failure. Metabolism *19:* 102–113 (1970).
29 Handelsman, D.: Hypothalamic-pituitary gondal dysfunction in renal failure, dialyses and renal transplantation. Endocr. Rev. *6:* 151–182 (1985).
30 Johnson, J.; Maack, T.: Renal extraction, filtration, absorption and catabolism of growth hormone. Am. J. Physiol. *233:* F 185 (1977).
31 Wright, A.; Lowy, G.; Fraser, T.; Spitz, I.; Rubenstein, A.; Bersonn, I.: Serum growth hormone and glucose intolerance in renal failure. Lancet *ii:* 798–800 (1968).
32 Phillips, I.; Fusco, A.; Untermen, T.; Greco, F. del: Somatomedin inhibitor in uremia. J. clin. Endocr. Metab. *59:* 764–772 (1984).
33 Drop, S.; Kortleve, D.; Guyda, H.; Posper, B.: Immunoassay of a somatomedin-binding protein from human amniotic fluid: levels in fetal, neonatal, and adult sera. J. clin. Endocr. Metab. *59:* 908–915 (1984).
34 Baxter, R.; Martin, J.: Radioimmunoessay of growth hormone-dependent insulin-like growth factor binding protein in human plasma. J. clin. Invest. *78:* 1504–1512 (1986).
35 Goldberg, A.; Trivedi, B.; Delmez, J.; Harter, H.; Daughaday, W.: Uremia reduces insulin-like growth factor I, increases insulin-like growth factor II, and modifies their serum protein binding. J. clin. Endocr. Metab. *55:* 1040–1045 (1982).
36 Powell, D.; Rosenfeld, R.; Baker, B.; Liu, F.; Hintz, R.: Serum somatomedin levels in adults with chronic renal failure the importance of measuring insulin-like growth factor I (IGF-I) and IGF-II in acid-chromatographed uremia serum. J. clin. Endocr. Metab. *63:* 1186–1192 (1986).
37 Finidori, J.; Postel-Vinay, C.; Kleinknecht, C.: Lactogenic and somatotropic binding

sites in liver membranes of rats with renal insufficiency. Endocrinology *106:* 1960 (1980).

38 Kinlaw, W.; Schwartz, H.L.; Mariash, C.N.; Bingham, C.; Carr, F.E.; Oppenheimer, J.H.: Hepatic messenger ribonucleic acid activity profiles in experimental azotemia in the rat. J. clin. Invest. *74:* 1934 (1984).

39 Mehls, O.; Ritz, E.; Hunziker, E.-B.; Eggli, P.; Heinrich, U.; Zapf, J.: Improvement of growth and food utilization by human recombinant growth hormone in uremia. Kidney int. *33:* 45–52 (1988).

Prof. Otto Mehls, MD, Division of Pediatric Nephrology, Department of Pediatrics, University of Heidelberg, Im Neuenheimer Feld 150, D-6900 Heidelberg (FRG)

Contr. Nephrol., vol. 64, pp. 43–48 (Karger, Basel 1988)

Does Lead Play a Role in the Development of Renal Insufficiency?

E. Ritz [a], *J. Mann* [a], *A. Wiecek* [b]

[a]Department Internal Medicine, University of Heidelberg, Heidelberg, FRG; [b]Silesian School of Medicine, Institute of Internal Medicine, Department of Nephrology, Katowice, Poland

In 1872, Lancereaux [1] first described contracted kidneys with non-specific vascular and interstitial sclerosis in workers who had a history of occupational lead exposure and lead intoxication. Such individuals had little or no proteinuria. It was subsequently recognized that lead intoxication was not only associated with renal failure but also with gout. Indeed, Wagner [2] who worked in Leipzig, the citadel of book printing, noted in the late 19th century that one case of gout was seen among 39 cases with chronic plumbism, but only one case among 1,490 patients without lead exposure. Among his patients with renal failure, no less than 15% had a history of lead intoxication.

At the turn of the centuries, *lead intoxication* was a well-known *cause of renal failure,* mentioned by Osler [3] or by Volhard and Fahr [4]. Because of improvements in industrial hygiene, this cause of renal failure virtually disappeared in Western countries. Many authors even questioned the relation of lead to renal failure.

For a long time the only convincing evidence for such interrelationship came from epidemiological observations, notably the Queensland epidemic [5] and the endemic in the moonshine whisky belt of the USA [6].

In Queensland (Australia) houses were perched on stilts and had lofty verandas the rails of which were so low that toddlers could conveniently lick rain drops left over from daily subtropical rain showers. Since the rails were painted with lead white, such rain drops had a sweet taste and children were thus exposed to very high doses of lead. Later they developed renal failure even after decades.

In contrast to acute lead intoxication (which is characterized by prominent inclusion bodies in the proximal tubular epithelial cells and acute proximal tubular dysfunction with Fanconi syndrome) such delayed saturnine renal failure is characterized by nonspecific clinical features: hypertension, shrunken kidneys with vascular sclerosis, glomerular obsolescence and interstitial fibrosis [7]. Of note, however, according to Emmerson [5], was the curious finding that over 50% of the Queensland patients with saturnine nephropathy suffered from gout. Thus, gout may be an indicator of chronic lead excess in uremic patients.

The interest in saturnine nephropathy was rekindled by Richet et al. [9, 10], who studied 23 workers who had a history of occupational lead exposure and developed renal failure, hypertension and – in a certain proportion – gout. Isolated cases of occupationally acquired lead nephropathy are infrequently seen in Western Europe [9, 12, 13], more so in industrially underdeveloped countries like Yugoslavia [13] or Rumania [14]. In the FRG, less than 200 individuals currently receive rent for diseases caused by occupational lead exposure [15]. The problem may be underestimated, however, since occupationally acquired lead nephropathy continues to be a problem in the USA [16–19].

Diagnosis of lead nephropathy in individuals with renal failure may be rendered difficult by the recent recognition that chronic renal failure from lead may occur despite the absence of a history of acute symptomatic lead intoxication. This is clearly illustrated by several recent catamnestic analyses of lead-exposed workers [18, 19]. In one study, 1,987 workers employed in an Idaho lead smelter between 1940 and 1965 were found to have excess mortality from chronic renal disease, standardized mortality ratio reaching 392 with prolonged duration of employment [19]. Cooper et al. [18] noted increased overall mortality, mortality from hypertension and chronic renal disease in almost 7,000 workers occupationally exposed to lead.

Apart from the relation between chronic renal disease and lead, the causative relationship between *lead and hypertension* has currently attracted considerable interest. In occupational lead exposure, a modest but significant risk of hypertension has recently been well documented. Kirkby and Gyntelberg [20] found small but significant elevation of supine diastolic blood pressure, 83 vs 78 mm Hg, in 96 Danish battery workers exposed to lead for 9–45 years. In a prospective study of 80

Boston policemen [21], blood lead levels at the beginning of the study were a significant predictor of the later increment in systolic blood pressure.

The data of the NHANES survey (National Health and Nutrition Examination Survey) recently documented a significant relation between blood lead levels on the one hand and systolic and diastolic blood pressure on the other in the general population [22, 23]. This relation remained significant when age, sex, body weight, nutritional habits, hematocrit and several other confounding variables were considered. In conjunction with previous less impressive epidemiological work, these data provide strong circumstantial evidence, though not definite proof, that lead is a causative factor for increased blood pressure in the general population. This raises concern that rising environmental lead burden may contribute to blood pressure evolution in the population.

In addition to the undoubted renal failure in individuals with occupational lead exposure [9–19], some recent work also suggests that elevated body *lead* burden may also be related to *progression of renal failure* in patients with preexistent renal disease [8, 22–24].

Behringer [8] examined patients with renal failure with or without gout, using the EDTA lead mobilization test according to Emmerson [5]. The increment in urinary lead over baseline is taken as an index of total body lead burden. Measurements established a very low upper limit of normal in the healthy German population. This was attributed to higher accuracy of currently employed methods. It came as a complete surprise that a considerable proportion of patients with known renal disease like glomerulonephritis or polycystic kidney disease had post-EDTA urinary lead excretion above the highest value in the control population. Recently (table I) we found that 13 of 43 uremic patients with various renal diseases had mobilizable lead above the highest value in controls.

This confirms previous suggestions of Emmerson [5] that patients with renal failure who developed gout when serum creatinine had already been elevated, almost invariably showed elevated lead burden. In contrast, patients who presented with gout, and later developed renal failure of unknown cause, almost uniformly failed to have an elevated lead burden (table II). This confirms, in agreement with our experience [8, 25], that coexistence of renal failure and de novo gout may be a clinically useful indicator of underlying chronic saturnism.

Several possible errors must be considered in the interpretation of the above data. The most serious one is that EDTA pharmacokinetics are

Table I. Urinary lead excretion after EDTA administration

	Post-EDTA lead excretion nmol/day/1.73 m^2
Controls (n = 21) 23–69 years	306 (131–1,587)
Chronic renal failure patients (n = 43) 22–69 years	882 (160–3,609)
S-creatinine above 1.5 mg/dl (in 14 of the 43 patients)	1,587 nmol/4 days/1.73 m^2

From Ritz et al. [25].

Table II. Lead mobilization in patients with gout and renal failure

	Post-EDTA lead excretion nmol/4 days/1.73 m^2
Controls (n = 19)	306 (131–1,587)
Impaired renal function prior to gout (n = 8)	1,827 (648–360)
Gout prior to impaired renal function (n = 8)	936 (438–2,261)

altered in renal failure because of prolonged half-life of EDTA. Consequently, skeletal lead might have a greater 'chance' of being chelated and excreted in the urine. Chelation may further be increased because of secondary hyperparathyroidism which increases the internal bone surfaces and the chances of chelation. Presumably this is not the only explanation since careful history revealed definite past lead exposure in 80% of our cases.

The observation of a high prevalence of unsuspected elevation of lead burden in patients with renal failure was also confirmed in a independent study with different methodology. Van de Vyver et al. [25, 27] showed excessive elevation of lead/calcium ratios in iliac bone biopsy samples of approximately 5% of dialysis patients who had been biopsied for reasons unrelated to lead.

Since elevated lead burden is common in patients with renal failure, the question arises whether elevated lead burden is causally related to progression of renal failure. Although coincidence does not proof causality, a pathogenetic relation is plausible on the basis of the following data.

First, although it is very difficult to produce renal damage in adult lead-exposed rats, lead exposure in the perinatal period causes marked inhibition of renal growth and renal failure [28]. This may be related to the actions of lead at the nuclear level. It is conceivable that lead interferes with renal compensatory hypertrophy and repair of renal damage. Of potentially even greater importance is the observation of Tange et al. [29] who failed to observe renal damage of lead exposure in intact rats, but definite focal renal scarring when unilaterally nephrectomized rats were exposed to lead. This may indicate that remnant kidneys are particularly susceptible to further lead-induced renal damage.

Finally, even minor elevations of blood pressure might play a role in the adverse effect of lead on the evolution of renal function in patients with preexistent renal disease [30].

These preliminary, but tantalizing observations raise the possibility that lead is a nonspecifically damaging factor promoting progression of renal failure, a concept which remains a challenge to be proven in the future.

References

1 Lancereaux, E.: Néphrite et arthrite saturnines: coincidence de ces affections: parallèle avec la néphrite et l'arthrite goutteuses. Archs gén. Méd. *6:* 641 (1881).

2 Wagner, E.: Handbuch der Krankheiten des Harnapparates (Erste Hälfte, Der Morbus Brightii), p. 291 (Vogel, 1882).

3 Osler, W.: The principles and practice of medicine; 6th ed., p. 375 (Appleton, London 1907).

4 Volhard, F.; Fahr, T.: Die Bright'sche Nierenkrankheit (Springer, Berlin 1914).

5 Emmerson, B.T.: Chronic lead nephropathy. Kidney int. *5:* 1–5 (1973).

6 Ball, G.V.; Morgan, J.M.: Chronic lead ingestion and gout. Sth. med. J. *61:* 21–24 (1968).

7 Ritz, E.; Mann, J.; Stoeppler, M.: Lead and the kidney; in Crosnier et al., Actual. Néphrol. Hôpital Necker, pp. 225–254 (Flammarion, Paris 1987).

8 Behringer, D.; Craswell, P.; Mohl, C.; Stoeppler, M.; Ritz, E.: Urinary lead excretion in uremic patients. Nephron *42:* 323–329 (1986).

9 Richet, G.; Albahary, C.; Morel-Maroger, L.; Guillaume, P.; Galle, P.: Les altérations rénales dans 23 cas de saturnisme professionel. Soc. Méd. Hôp. Paris *117:* 441 (1966).

10 Richet, P.; Mignon, G.F.; Ardaillou, R.: Goutte secondaire des nephropathies chroniques. Presse méd. *73:* 633 (1965).

11 Rossi, I.: La funzionalità renale nel saturnismo professionale. Rif. Med. *30:* 841–851 (1956).

12 Crepet, M.; Gobbato, F.; Martinio, P.; Siracusa, F.: La nefropatia saturnina. Minerva nefrol. *3:* 69 (1959).

13 Radosevic, Z.; Saric, M.; Beritic, T.; Knezevic, J.: The kidney in lead poisoning. Br. J. Ind. Med. *18:* 222 (1961).

14 Lilis, R.; Gavrilescu, N.; Nestrorescu, B.; Dumitriu, C.; Roventa, A.: Nephropathy in chronic lead poisoning. Br. J. Ind. Med. *25:* 169 (1968).

15 Lehnert, G.; Szadkowski, D.: Die Bleibelastung des Menschen (Verlag Chemie, Weinheim 1983).

16 Wedeen, R.P.; Mallik, D.K.; Batuman, V.: Detection and treatment of occupational lead nephropathy. Archs intern. Med. *139:* 53–57 (1979).

17 Hong, C.D.; Hanenson, I.B.; Lerner, S.; Hammond, P.B.; Pesce, A.J.; Pollak, V.E.: Occupational exposure to lead: effects on renal function. Kidney int. *18:* 489–494 (1980).

18 Cooper, W.C.; Wong, W.; Kheifets, L.: Mortality among employees of lead battery plants and lead-producing plants, 1947–1980. Scand. J. Work. envir. Hlth *11:* 331–345 (1985).

19 Selevan, S.G.; Landrigan, P.J.; Stern, F.B.; Jones, J.H.: Mortality of lead smelter workers. Am. J. Epidem. *122:* 673 (1985).

20 Kirkby, H.; Gyntelberg, F.: Blood pressure and other cardiovascular risk factors of long-term exposure to lead. Scand. J. Work. envir. Hlth *11:* 15–19 (1985).

21 Weiss, S.T.; Munoz, A.; Stein, A.; Sparrow, D.; Speizer, F.E.: The relationship of blood lead to blood pressure in a longitudinal study of working men. Am. J. Epidem. *123:* 800 (1986).

22 Pirkle, J.L.; Schwartz, J.; Landis, J.R.; Harlan, W.R.: The relationship between blood lead levels and blood pressure and its cardiovascular risk implications. Am. J. Epidem. *121:* 246–255 (1985).

23 Harlan, W.R.; Landis, J., Jr.; Schmouder, R.L.: Relationship of blood lead and blood pressure in the adolescent and adult US population. J. Am. med. Ass. *253:* 530–534 (1985).

24 Ritz, E.; Mann, J.; Nowack, R.: Lead and hypertension. 11. Wissenschaftliche Tagung der Internationalen Gesellschaft für Hypertonie, Heidelberg 1986.

25 Ritz, E.; De Broe, M.; Stoeppler, M.; Van de Vyver, F.L.; Koster, J.; D'Haese, P.C.; Wedeen, R.P.: Does lead contribute to deterioration of renal function in patients with renal disease? (Letter to the Editor). Nephrol. Dial. Transplant. *1:* 206–207 (1986).

26 Colleoni, N.; D'Amico, G.: Chronic lead accumulation as a possible cause of renal failure in gouty patients. Nephron *44:* 32–35 (1986).

27 Van de Vyver, F.L.; D'Haese, P.C.; Visser, W.J.; Elseviers, M.M.; Knippenberg, L.J.; Lamberts, L.V.; Wedeen, R.P.; De Broe, M.E.: Bone lead in dialysis patients. Kidney int. *33:* 601–608 (1988).

28 Aviv, A.; John, E.; Bernstein, J.; Goldsmith, D.I.; Spitzer, A.: Lead intoxication during development: its late effects on kidney function and blood pressure. Kidney int. *17:* 430 (1980).

29 Tange, J.D.; Hayward, N.J.; Brenner, D.A.: Renal lesions in experimental plumbism and their clinical implications. Australas. Ann. Med. *14:* 49–56 (1965).

30 Ritz, E.; Wiecek, A.; Mann, J.: Nierenfunktion bei Bleibelastung. Klin. Wschr. *64:* 871–875 (1986).

Prof. E. Ritz, MD, Department of Internal Medicine, University of Heidelberg, Bergheimerstrasse 56, D-6900 Heidelberg (FRG)

Contr. Nephrol., vol. 64, pp. 49–57 (Karger, Basel 1988)

Bone Biopsy Studies in the Diagnosis and Treatment of Renal Osteodystrophy

J.L. Sebert [a], *P. Fardellone* [a], *C. Noel* [b], *M. Cohen-Solal* [a], *R. Bellony* [a], *A. Marie* [a], *A. Fournier* [a]

[a]Service de Nephrologie et Service de Rhumatologie, Centre Hospitalier Universitaire, Amiens; [b]Service de Nephrologie, Centre Hospitalier Universitaire, Lille, France

Bone biopsy has been used in patients with renal failure for diagnostic purposes and to assess the response to treatment. Since bone tissue is very sensitive to the metabolic disorders induced by renal insufficiency, histological bone lesions are already present in early chronic renal insufficiency and nearly all patients with terminal renal failure have histological bone disease [Malluche et al., 1976]. Bone biopsy is a unique procedure to evaluate bone disease and has provided major contributions to the identification, classification and pathophysiology of renal osteodystrophies [Sherrard et al., 1974, 1983; Teitelbaum, 1984].

We will first consider the methodological and general features of bone biopsy and bone histomorphometry, and secondly we will discuss the diagnostic value and the indications of bone biopsy in uremic patients.

Bone Biopsy and Bone Histomorphometry: Methodological Considerations

Bone Biopsy Procedure. The prerequisite for histological evaluation of bone is to obtain an intact bone sample. The most widely used approach is the transiliac technique. The sampling site is a 5-cm isolateral triangular area (Bordier's triangle) located behind the anterior superior iliac spine and below the iliac crest border. A manual trephine, 7 mm inner diameter, is generally used and provides an adequate amount of both cortical and trabecular bone without major distorsion of bone architecture. The procedure can be carried out in outpatients under local anesthesia. During sampling, the pressure applied to the instrument must be as gentle as possible in order to avoid the fragmentation of the bone specimen. Since trabecular bone mass is generally not reduced in renal osteodystrophy, the quality of bone samples is far better in uremic patients than in patients with other metabolic bone disease. The transiliac biopsy can be repeated on the opposite side for following-up patients. A time interval of about one year is required between 2 sampling procedures on the same iliac side to avoid the local acceleration of bone turnover induced by the repair process of the earlier biopsy.

The safety of transiliac bone biopsy has been examined on large series of patients reviewed by Rao [1983]. Complications occurred in 0.7% of 9,131 patients and consisted mainly of local hematoma and pain lasting more than a week. In our experience these complications occurred in about 4.5% (7 cases) of 150 biopsies performed in hemodialyzed patients. This higher frequency could be explained by a higher hemorrhagic risk in patients on regular hemodialysis. In these patients, bone biopsy should be performed on an interdialysis day. In addition, in patients with a prolonged bleeding time, intranasal Desmopressin (DDAVP) could be of use to decrease the hemorrhagic hazards [Nald et al., 1985].

Bone Histomorphometric Parameters. Complete histological evaluation of bone biopsies requires that bone sections are observed without previous decalcification of the specimen, in order to preserve the distinction between calcified and uncalcified (osteoid) bone and to allow the study of the mineralization process. This needs embedding of the bone specimen in plastic medium such as methylmethacrylate [Baron, 1983].

The histomorphometric analysis of trabecular bone includes the determination of: (1) parameters of bone mass (trabecular bone volume and osteoid volume); (2) static parameters of bone formation and resorption (osteoid and osteoblastic surfaces, osteoid seam width, resorption and osteoclastic surfaces, osteoclast count), and (3) dynamic parameters of bone formation and mineralization (tetracycline labelled surfaces, mineral appositional rate, bone formation rate). The study of these dynamic parameters of bone remodelling is based on the binding of tetracycline antibiotics to actively mineralizing bone surfaces and requires that an in vivo double tetracycline labelling has been performed before the biopsy procedure [Rasmussen and Bordier, 1974; Melsen and Mosekilde, 1978; Meunier and Bressot, 1982; Parfitt, 1983; Malluche and Faugere, 1986].

The reproducibility of bone histomorphometric measurements has been studied on paired biopsies obtained from both normal and uremic patients [De Vernejoul et al., 1981]. As shown in table I, the overall variability is lower in uremic patients than in normal subjects. However, the intraindividual variation in uremic patients is sometimes large, especially for parameters of bone resorption. This limits the value of sequential biopsies in an individual patient. For example, a minimal variation of more than 50% is required to ascertain that a change in bone resorption is significant and not only due to inherent variability.

Histological Evaluation of Renal Osteodystrophy and Indication for Bone Biopsy

The bone histology of uremic patients has been studied extensively [Ellis and Peart, 1973; Sherrard et al., 1974; Malluche et al., 1976; Evans et al., 1982]. Besides hyperparathyroidism and osteomalacia, other types of

Table I. Bone histomorphometry: reproducibility in normal and uremic patients [adapted from de Vernejoul et al., 1981]

	Intraindividual variation, %				Minimal significant variation, % (uremic patients)
	normal patients		uremic patients		
	mean	SD	mean	SD	
Trabecular bone volume	19.1	13.8	16.9	14.1	45
Osteoid volume	45.5	31.2	18.2	14.3	46
Osteoid surface	34.7	29.4	9	12.8	35
Osteoid thickness	–	–	3.5	2.6	9
Resorption surface	52.3	44	23.7	17.4	59
Osteoclast count	–	–	21.1	11.7	47
Mineral appositional rate	–	–	5.9	4.3	15

Table II. Histological classification of renal osteodystrophy [according to Sherrard et al., 1983]

Type of bone disease	Marrow fibrosis % bone tissue area	Osteoid volume % trabecular bone volume	Bone formation rate % of normal
Osteitis fibrosa	>0.5	<15	>100
Mild	<0.5	<15	> 30
Osteomalacia	<0.5	>15	< 30
Mixed	>0.5	>15	< 30
Aplastic	<0.5	<15	< 30

bone disease have more recently enlarged the spectrum of renal osteodystrophy [Sherrard et al., 1983]. When both static and dynamic parameters of bone histomorphometry are considered, patients should be classified into 5 different groups (table II). Osteitis fibrosa and mild bone disease reflect various degrees of hyperparathyroidism and are associated with high bone turnover and high bone formation rate. On the contrary, bone formation rate is decreased in osteomalacia and aplastic bone disease. Figure 1 shows the wide spectrum of individual values for bone resorption and bone formation surfaces in uremic patients depending on their bone turnover level.

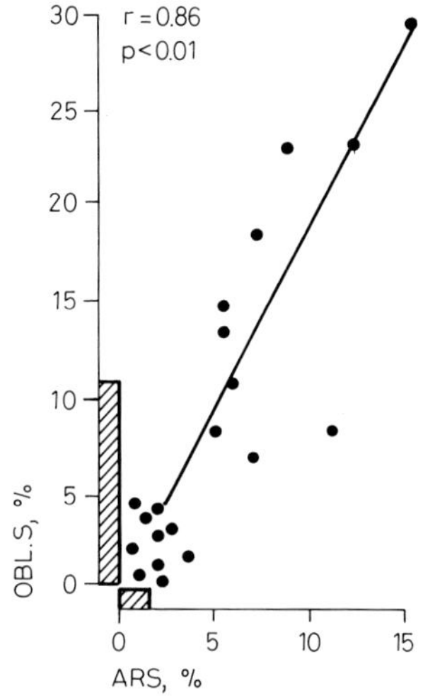

Fig. 1. Correlation between osteoblastic surfaces (OBL.S) and active resorption surfaces (ARS) in 20 uremic patients on chronic hemodialysis.

Predominant hyperparathyroidism is the most frequent bone lesion and is histologically found in about 50% of unselected asymptomatic patients on chronic hemodialysis. It can progress to radiological and symptomatic bone disease. We examined the prevalence of radiological hyperparathyroidism in 750 hemodialyzed patients from the North of France and found that 12% had subperiosteal erosions on hand X-rays (fig. 2); 5% required parathyroidectomy. A higher incidence of hyperparathyroidism was noted in home-dialyzed patients, presumably because of a lower compliance to treatment.

The high bone turnover which characterizes hyperparathyroid bone disease can be predicted by plasma analysis. In addition to measurements of parathyroid hormone and alkaline phosphatase, the diagnostic value of plasma bone Gla-protein has been recently emphasized in renal osteodystrophy [Malluche et al., 1984; Charhon et al., 1986a, b]. Bone Gla-protein is a non collagenic bone protein which is synthesized by osteoblasts, and its plasma level is a sensitive marker of bone formation and bone turnover.

We have recently examined the relative values of plasma bone Gla-protein, parathyroid hormone and alkaline phosphatase for the prediction of bone formation status in uremic patients. As shown in table III,

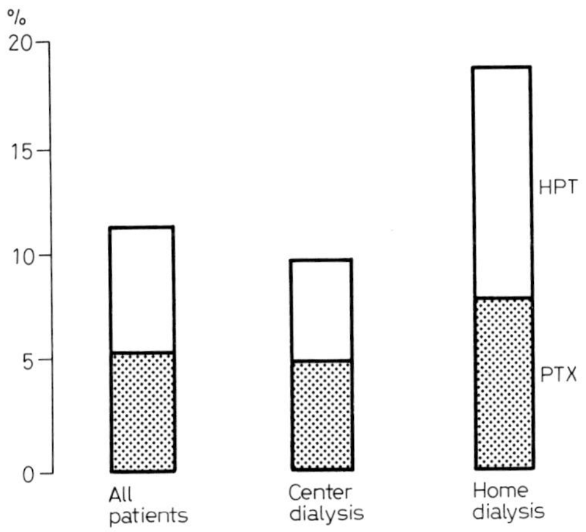

Fig. 2. Prevalence of radiological hyperparathyroidism (HPT) and of parathyroidectomy (PTX) in 764 patients on chronic hemodialysis.

these 3 biochemical markers were correlated with the histomorphometric parameters of bone turnover and bone formation. When patients were divided into 2 groups according to their bone formation rate (table IV), bone Gla-protein was significantly higher in the group with high bone formation rate (predominant hyperparathyroid bone disease) than in the group with low bone formation rate. However, alkaline phosphatase was not significantly different in the 2 groups. When the predictive value was assessed in terms of sensitivity and specificity, plasma bone Gla-protein was a better index of bone formation than plasma parathyroid hormone.

Osteomalacia and aplastic bone disease are associated with low bone formation rates. Osteomalacia can occur in association with hyperparathyroidism (mixed bone disease). This is most frequently seen in undialyzed uremic patients, and contributing factors such as vitamin D deficiency and acidosis should be considered in the pathogenesis of the mineralization defect [Eastwood et al., 1977]. Since these patients also have severe secondary hyperparathyroidism the differential diagnosis between pure osteitis fibrosa and mixed bone diseasc may be difficult to ascertain on a clinical or biological basis and requires histological confirmation.

Table III. Correlations between bone histomorphometric parameters and plasma bone Gla-protein (BGP), alkaline phosphatase (Alk. Pase) and parathyroid hormone (PTH)

Histomorphometric parameters	Correlation coefficient (r) with		
	BGP	Alk.Pase	PTH
Osteoid volume	0.46*	0.42	0.17
Osteoid surface	0.36	0.27	0.008
Osteoblastic surface	0.78***	0.68***	0.64***
Active resorption surface	0.54**	0.61***	0.58***
Osteoclast count	0.54**	0.51**	0.58**
Total labelled surface	0.75***	0.56**	0.44*
Mineral appositional rate	0.47*	0.48*	0.49*
Bone formation rate	0.76***	0.55**	0.51**

* $p<0.05$; ** $p<0.01$; *** $p<0.001$.

Table IV. Comparison of histomorphometric and biochemical results (mean ± SE) between 2 groups of patients classified according to their bone formation rate

	Bone formation rate		Normal values mean ± SD or range
	low (n = 17)	high (n = 10)	
Histomorphometry			
Trabecular bone volume	21.6 ± 1.4	24.4 ± 2.5	21 ± 5
Osteoid volume	6.1 ± 0.9	7.4 ± 1	2.5 ± 1
Osteoid surface	47.7 ± 3.9	47.9 ± 4.7	16 ± 6
Osteoblastic surface	4.6 ± 0.9	18.7 ± 2.8***	5 ± 3
Active resorption surface	3.1 ± 0.7	9 ± 1.2***	0.5 ± 0.5
Osteoclast count	0.59 ± 0.14	2.1 ± 0.3***	0.2 ± 0.1
Total labelled surface	4.9 ± 0.9	26 ± 2.9***	14 ± 1.3
Biochemistry			
Calcium, mmol/l	2.4 ± 0.05	2.36 ± 0.04	2.10 – 2.62
Phosphate, mmol/l	1.8 ± 0.1	1.97 ± 0.14	0.8 – 1.45
Aluminum, µmol/l	2.01 ± 0.57	1.41 ± 0.64	< 0.30
PTH, pg/ml	176 ± 30	305 ± 45*	20 – 45
Alk.Pase, IU/l	129 ± 15	200 ± 36	60 – 170
BGP, ng/ml	14.8 ± 3	58 ± 13***	2 – 10

Student's t-test for unpaired data: * $p<0.05$; *** $p<0.001$.

In hemodialyzed patients, aluminum intoxication is the major factor leading to osteomalacia and aplastic bone disease. These aluminum-related osteopathies are often symptomatic and have mainly been reported in heavily intoxicated patients, frequently in association with encephalopathy and hypercalcemia [Parkinson et al., 1979; Hodsman et al., 1982; Norris et al., 1985]. The accumulation of aluminum in bone can be demonstrated on undecalcified bone sections using a simple histochemical method [Maloney et al., 1982]. These histochemical studies have shown that aluminum is mainly located at the interface between mineralized bone and osteoid, i.e. the mineralization front. Aluminum is also found inside bone along the cement lines. The histochemical method is well correlated with the direct measurement of bone aluminum by atomic absorption spectrometry. Whereas plasma concentration of aluminum in the basal state is not a reliable indicator of bone aluminum overload, the rise of plasma aluminum after infusion of desferrioxamine is better correlated with the aluminum bone content and the presence of osteomalacia or aplastic bone disease [Milliner et al., 1984]. However, bone aluminum overload with or without mineralization defect can occur in patients with predominantly hyperparathyroid bone disease [Charhon 1986a, b]. An acceleration of aluminum deposition in bone has been reported after parathyroidectomy [Andress et al., 1985; Charhon et al., 1985, De Vernejoul et al., 1985]. This could play a role in the emergence of aplastic bone disease after parathyroidectomy.

Since no indirect biochemical method can, at present, adequately predict the degree of bone aluminum overload, it seems advisable to perform a bone biopsy in individual patients who require a major treatment such as parathyroidectomy. It seems also reasonable to repeat biopsies during the follow-up of these patients in order to detect changes in bone histomorphometric profiles and in aluminum content. In addition, a biopsy should be performed systematically in patients with hypercalcemia, since different types of bone disease, including hyperparathyroidism and aluminum-related osteopathies, can be observed in this situation.

References

Andress, D.L.; Ott, S.M.; Maloney, N.A.; Sherrard, D.J.: Effect of parathyroidectomy on bone aluminum accumulation in chronic renal failure. New Engl. J. Med. *312:* 468–473 (1985).

Baron, R.; Vignery, A.; Neff, L.; Silvergate, A.; Santa Maria, A.: Processing of undecalcified bone specimens for bone histomorphometry; in Recker, Bone histomorphometry: techniques and interpretation, pp. 13–35 (CRC Press, Boca Raton 1983).

Charhon, S.A.; Berland, Y.F.; Olmer, M.J.; Delawari, E.; Traeger, J.; Meunier, P.J.: Effects of parathyroidectomy on bone formation and mineralization in hemodialyzed patients. Kidney int. *27:* 426–435 (1985).

Charhon, S.A.; Chavassieux, P.M.; Chapuy, M.C.; Traeger, J.; Meunier, P.J.: High bone turnover associated with an aluminum-induced impairment of mineralization. Bone *7:* 319–324 (1986a).

Charhon, S.A.; Delmas, P.D.; Malaval, L.; Chavassieux, P.M.; Arlot, M.; Chapuy, M.C.; Meunier, P.J.: Serum bone Gla-protein in renal osteodystrophy: comparison with bone histomorphometry. J. clin. Endocr. Metab. *63:* 892–897 (1986b).

Eastwood, J.B.; Stamp, T.C.B.; De Wardener, H.E.; Bordier, P.J.; Arnaud, C.D.: The effect of 25 hydroxy vitamin D_3 in the osteomalacia of chronic renal failure. Clin. Sci. mol. Med. *52:* 499–508 (1977).

Ellis, H.A.; Peart, K.M.: Azotaemic osteodystrophy: a quantitative study on iliac bone. J. clin. Path. *26:* 83–101 (1973).

Evans, R.A.; Flynn, J.; Dunstan, C.R.; George, C.R.P.; MacDonnel, G.D.: Bone metabolism in chronic renal failure. Mineral Electrolyte Metab. *7:* 207–218 (1982).

Hodsman, A.B.; Sherrard, D.J.; Alfrey, A.C.; Ott, S.; Brickman, A.S.; Miller, N.L.; Maloney, N.A.; Coburn, J.W.: Bone aluminum and histomorphometric features of renal osteodystrophy. J. clin. Endocr. Metab. *54:* 539–546 (1982).

Malluche, H.H.; Faugere, M.C.; Fanti, P.; Price, P.A.: Plasma levels of bone Gla-protein reflect bone formation in patients on chronic maintenance dialysis. Kidney int. *26:* 869–874 (1984).

Malluche, H.H.; Faugere, M.C.: Atlas of mineralized bone histology (Karger, Basel 1986).

Malluche, H.H.; Ritz, E.; Lange, P.H.; Kutschera, J.; Hodgson, M.; Seiffert, U.; Schoeppe, W.: Bone histology in incipient and advanced renal failure. Kidney int. *9:* 355–362 (1976).

Maloney, N.A.; Ott, S.; Aflrey, A.C.; Miller, N.L.; Coburn, J.W.; Sherrard, D.J.: Histological quantitation of aluminum in iliac bone from patients with renal failure. J. Lab. clin. Med. *99:* 206–216 (1982).

Melsen, F.; Mosekilde, L.: Tetracycline double-labelling of iliac trabecular bone in 41 normal adults. Calcif. Tissue Res. *26:* 99–102 (1978).

Meunier, P.J.; Bressot, C.: Endocrine influences on bone cells and bone remodelling evaluated by clinical histomorphometry; in Parsons, Endocrinology of calcium metabolism, pp. 445–465 (Raven Press, New York 1982).

Milliner, D.S.; Nebeker, H.G.; Ott, S.M.; Andress, D.L.; Sherrard, D.J.; Alfrey, A.C.; Slatopolsky, E.A.; Coburn, J.W.: Use of the desferrioxamine infusion test in the diagnosis of aluminum-related osteodystrophy. Ann. intern. Med. *101:* 775–780 (1984).

Nald, C.; Richardson, W.; Robinson, A.G.: Desmopressin. Ann. intern. Med. *103:* 228–239 (1985).

Norris, K.C.; Crooks, P.W.; Nebeker, H.G.; Hercz, G.; Milliner, D.S.; Gerszi, K.; Slatopolsky, E.; Andress, D.L.; Sherrard, D.J.; Coburn, J.W.: Clinical and laboratory features of aluminum-related bone disease: differences between sporadic and 'epidemic' forms of the syndrome. Am. J. Kidney Dis. *6:* 342–347 (1985).

Parfitt, A.M.: The physiological and clinical significance of bone histomorphometric data; in Recker, Bone histomorphometry: techniques and interpretation, pp. 143–223 (CRC Press, Boca Raton 1983).
Parkinson, I.S.; Ward, M.K.; Feest, T.G.; Fawcett, R.W.P.; Kerr, D.N.S.: Fracturing dialysis osteodystrophy and dialysis encephalopathy. An epidemiological survey. Lancet *i:* 406–409 (1979).
Rao, D.S.: Practical approach to bone biopsy; in Recker, Bone histomorphometry: techniques and interpretation, pp. 3–11 (CRC Press, Boca Raton 1983).
Rasmussen, H.; Bordier, P.: The physiological and cellular basis of metabolic bone disease (Williams & Wilkins, Baltimore 1974).
Sherrard, D.J.; Baylink, D.J.; Wergedal, J.E.; Maloney, N.A.: Quantitative histological studies on the pathogenesis of uremic bone disease. J. clin. Endocr. Metab. *39:* 119–135 (1974).
Sherrard, D.; Ott, S.; Maloney, N.; Andress, D.; Coburn, J.: Uremic osteodystrophy: classification, cause and treatment; in Frame, Clinical disorders of bone and mineral metabolism, pp. 254–259 (Excerpta Medica, Amsterdam 1983).
Teitelbaum, S.L.: Renal osteodystrophy. Human Pathol. *15:* 306–323 (1984).
Vernejoul, M.C. de; Kuntz, D.; Miravet, L.; Gueris, J.; Bielakoff, J.; Ryckewaert, A.: Bone histomorphometry in hemodialysed patients. Metab. Bone Dis. Rel. Res. *3:* 175–179 (1981).
Vernejoul, M.C. de; Marchais, S.; London, G.; Morieux, C.; Bielakoff, J.; Miravet, L.: Increased bone aluminum deposition after subtotal parathyroidectomy in dialyzed patients. Kidney int. *27:* 785–791 (1985).

J.L. Sebert, MD, Service de Rhumatologie, Hôpital Nord, Centre Hospitalier Universitaire, F–80030 Amiens (France)

Contr. Nephrol., vol. 64, pp. 58–73 (Karger, Basel 1988)

Calcium Carbonate and Magnesium Hydroxide in the Prevention of Renal Osteodystrophy or the Demise of Aluminum Toxicity in Uremia

Analysis of 5 Years Experience

P. Morinière[a], *A. Fournier*[a], *P.F. Westeel*[a], *A. Idrissi*[a], *H. Renaud*[a], *C. Hocine*[a], *S. Belbrik*[a], *A. Marie*[b], *P. Leflon*[c], *J.L. Sebert*[d]

[a]Service de Néphrologie, [b]Laboratoire d'Anatomie pathologique, [c]Laboratoire de Biochimie, [d]Service de Rhumatologie, CHU d'Amiens, France

Aluminum intoxication in uremic patients may be due not only to dialysate contamination but also to the absorption of aluminum from oral phosphate binders. Therefore, alternatives to these binders are highly desirable. High doses of calcium carbonate ($CaCO_3$) and magnesium dihydroxide or carbonate have been proposed with that purpose.

The aim of this paper is to review the rationale of their use and their effectiveness and safety on the basis of our own experience and that of the literature.

Rationale for the Use of High Doses of $CaCO_3$ in Uremia

Renal osteodystrophy is a complex metabolic disease usually associating various degrees of increased bone resorption related to secondary hyperparathyroidism associated with a mineralization defect resulting in florid osteomalacia. More rarely an aplastic bone disease is present in dialysis patients eventually resulting in osteopenia. Hyperparathyroidism is mainly triggered [1, 2] by the tendency of plasma calcium to decrease in renal failure. This phenomenon is the result of 2 main factors: the hyperphosphatemia secondary to the decrease of glomerular filtration rate (GFR) and the negative calcium balance secondary to dietary restriction in milk and meat [2] and to the decreased synthesis of 1α-$25(OH)_2D_3$ the active metabolite of vitamin D_3 which does not appropriately stimulate the calcium absorption when the calcium intake is low (<1 g/day). The decreased synthesis of 1α-$25(OH)_2D_3$ is due to two main factors: (a) the decreased nephron mass, (b) the increased phosphate concentrations in the blood and the reduction of tubular cell 1α-hydroxylase activity and, sometimes, to vitamin D depletion especially in northern industrialized countries where food is not fortified with vitamin D because of lack of sun, fat malabsorption or anticonvulsants. In dialyzed patients negative calcium balance may also be induced by a too low dialysate calcium (<6.0 mg/l). Finally, hyperstimulation of parathyroid hormone (PTH) secretion may be due to the upward shift of the set point for calcium in which not only the decreased plasma levels of $1,25(OH)_2D_3$ and/or $24,25(OH)_2D_3$ [1] but also phosphate retention per se play independent roles [3].

The pathophysiology of uremic osteomalacia is more elusive. Its occurrence is inconsistent especially in dialysis patients. No histological osteomalacia has been found in anephric patients not receiving vitamin D derivatives demonstrating that normal levels of 1α-$25(OH)_2D_3$ are not necessary to prevent osteomalacia [4]. On the contrary, epidemics of fracturing osteomalacia usually associated with a fatal encephalopathy have been reported and have been related to the high content of dialysate in aluminum [5]. However, in spite of permanent low dialysate aluminum occasional reports of aluminum bone disease have been published in dialysis patients. In a few cases, the bone disease was associated with encephalopathy and the disease has been related to aluminum overload slowly induced by aluminum absorption from the aluminum-containing phosphate binders [6]. In patients not yet on dialysis, true osteomalacia with increased osteoid seams and decreased mineralization rate has rarely been reported. In quite a number of cases physiological doses of $25(OH)D_3$ cured this osteomalacia which was therefore considered to be due to vitamin D depletion [7]. In other anecdotal cases it was due to calcium or phosphate deficiency or to anticonvulsant treatment [1]. More frequently it was related to aluminum intoxication from phosphate binders and usually associated to encephalopathy especially in children [6]. Long-standing acidosis particular to renal diseases involving predominantly the medulla like analgesic nephropathy has also been recognized as a risk factor of osteomalacia in these patients, as well as the low plasma levels of $1,25(OH)_2D_3$ responsible for the negative dietary calcium balance [6]. However, when only the latter factor is involved, the mineralization defect is usually characterized by a decreased mineral apposition rate without increased thickness of the osteoid seam, i.e. without florid osteomalacia [9].

The high suspicion of aluminum toxicity from the aluminum phosphate binders led us to study the evolution of plasma aluminum of our other patients after the installation of reverse osmosis in our center to remove aluminum from the tap water. We were surprised not to observe a significant fall in their plasma aluminum [10]. Furthermore, a significant correlation between the cumulative dose of $Al(OH)_3$ and not only the plasma concentration of aluminum but also the increase of plasma aluminum after desferrioxamine and the ponderal aluminum content of the bone, further increased our suspicion that the aluminum-containing phosphate binders were the source of the aluminum overload of our patients [11]. This led us to look for alternative phosphate binders. We remembered then our readings of 1968 when we were planning the studies which demonstrated in dialyzed men that plasma PTH levels were decreased by increasing the dialysate calcium from 5.5 to 6.5 mg/dl and by decreasing the plasma phosphate to less than 6.0 mg/dl by the use of $Al(OH)_3$ [12] and led us to recommend the general use of aluminum phosphate binders to control hyperparathyroidism. We reread then the pioneer work of Clarkson et al. [13] who showed, with balance studies, that 20 g of $CaCO_3$ in uremic men not yet on dialysis could induce a positive calcium balance with an increase in plasma calcium and a decrease of plasma phosphate. This decrease in plasma phosphate could be explained however not only by an increased fecal loss of phosphate but also by calcium phosphate deposition since the phosphate balance was actually positive, the decrease in phosphaturia counterbalancing the increased fecal loss. Where was the calcium phosphate deposited: in the bone or in the soft tissues? The possibility that the calcium phosphate deposition could well take place in bone was first suggested by the report of Curtis et al. [14] of the improvement of uremic rickets by high doses of calcium carbonate without occurrence of soft tissue calcifications. This suggestion was also pointed out by Meyrier et al. [15]. Although these authors used high doses of $CaCO_3$ (5–20 g/day) not with the aim to bind phosphate ($Al[OH]_3$ being used for that purpose at a daily dose of 2–5 g/day) but to increase calcium absorption, they found compared to a group taking no $Al(OH)_3$ and no calcium supplement, a favorable effect of this treatment on plasma levels of phosphate and alkaline phosphatases which were lower and on bone X-rays and biopsies which showed improvement of resorption and of osteomalacia. Furthermore, pseudogout attacks and metastatic calcifications were less frequent in this group (2/31 versus 9/24). Finally we were reassured by our own experience reported in the thesis of Roussel [16] about preventive treatment of osteodystrophy only by $Al(OH)_3$ (=3 g/day) and moderate doses of $CaCO_3$ (ranging from 0 to 6 g/day). There was no difference in doses of $CaCO_3$ between the patients who developed soft tissue calcifications and those who did not. Furthermore the speed of extension of the soft tissue calcifications was favored by age, male sex, the diastolic blood pressure, and blood levels of triglyceride, glucose, calcium and phosphate but not by the doses of $CaCO_3$ or of $Al(OH)_3$. Thus, $CaCO_3$ appears to favor calcification only by inducing hypercalcemia and not directly. Therefore, the hypothesis that the decrease of plasma phosphate was due to deposition of calcium phosphate in the soft tissues in proportion to the quantity of calcium absorbed and therefore to the dose of $CaCO_3$ administered, could be rejected.

Another reason to use $CaCO_3$ in uremia was its potential alkalinizing effect demonstrated by Clarkson et al. [13] in uremic patients not yet on dialysis. They observed an increase in plasma bicarbonate with an increase in urine pH, a decrease in residual titrable acidity and a decrease in NH_3 excretion. Therefore, the alkalinizing effect is not due to a renal generation of bicarbonates. The latter comes from the digestive tract according to the following reactions: – in the upper intestinal tract, $CaCO_3$ binds to 2 HCl to yield $CaCl_2 + H_2O + CO_2$; – in the lower intestinal tract $CaCl_2$ is either reabsorbed without modification or retransformed into $CaCO_3$ and eliminated in the feces after recombination with $NaHCO_3$. The $CaCl_2$ which is reabsorbed comes finally from the HCl neutralization which has generated HCO_3 in the plasma. The increased plasma levels of bicarbonate depress the excretion of H^+ by the kidney. The use of $CaCO_3$ as alkalinizing agent was proposed by Makof et al. [18] in 1969. More recently, a dramatic alkalinizing effect of $CaCO_3$ in dialyzed patients was reported by Papadoyannakis et al. [19] who showed that 2–4 g of $CaCO_3$ could increase plasma HCO_3^- from 16 to 23 mEq/l and decrease the protein catabolic rate from 1.2 to 1 g/kg/24 h.

Modality of Prescription

Since the use of high doses of $CaCO_3$ is aimed mainly at binding phosphate and not at increasing calcium absorption, $CaCO_3$ should be given with the meals rich in phosphate. In France, lunch and dinner represent the two main and equal intakes of phosphate. Therefore, we recommend to take $CaCO_3$ only and at equal doses at lunch and dinner.

To avoid soft tissue calcification by giving $CaCO_3$ while the plasma levels of phosphate are quite above 6.0 mg/dl, we prefer to give high doses of $CaCO_3$ once plasma phosphate has been decreased below 6.0 mg/dl by temporary administration of $Al(OH)_3$. $CaCO_3$ is then given at the daily dose of 3 g and then progressively increased while $Al(OH)_3$ is progressively discontinued.

Efficiency and Safety of High Doses of $CaCO_3$ as Phosphate Binder in Patients on Dialysis

Efficiency

This efficiency was first reported by Morinière et al. [20] in 1982 on a group of 23 patients (16 without and 7 with previous Vitamin D treatment). Previous treatment by $Al(OH)_3$ at a mean daily dose of 4.8 g was discontinued while the daily dose of $CaCO_3$ was increased from 5 to 12 g/day. After 6 months plasma phosphate were better controlled (5.0 versus 5.5 mg/dl) whereas alkaline phosphatase remained normal and plasma levels of C-terminal PTH stable. Furthermore, plasma aluminum decreased significantly from 58 to 32 μg/l demonstrating that the increased levels of plasma aluminum were due to $Al(OH)_3$ ingestion. A second study by Morinière et al. [21] not only confirmed this efficiency of $CaCO_3$ as phosphate binder but demonstrated in a randomized controlled study that the control of hyperparathyroidism (assessed PTH levels) was as good with high doses of $CaCO_3$ as with a combination of 3 g of $CaCO_3$, $Al(OH)_3$ and 1α-$(OH)D_3$ at doses giving the same levels of plasma phosphate and calcium. This study demonstrated therefore that the lack of high levels of $1,25(OH)_2D_3$ is not as deleterious for the control of hyperparathyroidism as suggested by some authors emphasizing the role of low levels of $1,25(OH)_2D_3$ on the increase of the set point of calcium. This observation is in agreement with the opinion of Slatopolsky et al. [3] who stress the importance of plasma phosphate levels per se in the upward shift of the set point of calcium.

In both works of Morinière et al. [20, 21] the alkalinizing effect of high doses of $CaCO_3$ was not evident since the plasma carbonate increased only by 1–2 mEq/l. The baseline values were, however, not as low (22 mEq/l) as in Papadoyannakis' experience.

Since these original works relating a cumulative experience of 18 months, numerous reports have been published confirming the efficiency of high doses of calcium as phosphate binder [22–25].

The Limits of $CaCO_3$ as Phosphate Binder

Clinical Side Effects. A few patients do not tolerate high doses of $CaCO_3$ because of diarrhea or constipation. In our original work diarrhea led to $CaCO_3$ discontinuation in 2 cases. The second side effect is a distaste that the patient develops after long-term treatment. To prevent this a change of commercial preparations is helpful: these include capsules, granules (Calcileve in sachets of 5 g – Le Franc Laboratoires, Paris), effervescent tablets (Calcium Sandoz Forte), drinkable vials (Calciforte, Serozym Laboratoires, Paris), orange flavored suspension (Laboratoire Valpan, Paris).

Biochemical Side Effects. Hypercalcemia is the principal side effect of high doses of $CaCO_3$. In the second controlled study the episodes of hypercalcemia (>11.0 mg/dl) were as frequent in the $CaCO_3$ groups as in the group treated with 1α-$(OH)D_3$ and $Al(OH)_3$ (22 episodes in 8 patients of the 15 $CaCO_3$ patients versus 25 episodes in 9 of the 12 patients treated with 1α-$(OH)D_3$). The occurrence of hypercalcemia is unpredictable and may appear when the dose of $CaCO_3$ has been stable for a long time. This justifies weekly monitoring of plasma calcium. Although unpredictable, hypercalcemia occurs mainly in the same patient. The patients with existing aluminum bone disease are more likely to develop hypercalcemia [24]. Treatment with desferrioxamine may increase the tolerance of $CaCO_3$ as we have found in one of our patients [6].

In order to increase the calcium tolerance we recommend [6] that the calcium balance of the renal function replacement procedure be no longer positive but null, i.e. for hemodialysis or peritoneal dialysis that the dialysate calcium be 6.0 mg/dl and for hemofiltration with 30 liters of ultrafiltration and a weight loss of 3 kg that the calcium concentration of the substitution fluid be 6.6 mg/dl. Furthermore, we do not advise in adults the systematic administration of 1α-hydroxyvitamin D since it increases calcium and phosphate absorption [26]. We recommend only the prevention of vitamin D depletion by vitamin D_2 or D_3 at a dose of

800 IU/day or by $25(OH)D_3$ at a weekly dose of 75 μg. In children small doses of $1,25(OH)_2D_3$ (0.25 to 0.75 μg/day) may be necessary and have been administered in association with high doses of $CaCO_3$ without frequent hypercalcemia [23].

Hypercalcemia can be managed by temporary discontinuation of $CaCO_3$ or dose reduction. This leads to recurrent hyperphosphatemia and when the above procedures have failed to improve the tolerance of $CaCO_3$, the use of other phosphate binders such as $Mg(OH)_2$ or Mg carbonate may be attempted when severe dietary phosphate restrictions and acceptable longer duration of dialysis have failed to control hyperphosphatemia. The concentration of calcium in the dialysate may also be decreased below 6.0 mg/dl to even 4.0 mg/dl.

We have found only a limited alkalinizing effect of $CaCO_3$ in hemodialyzed patients. In order to normalize the plasma bicarbonate, we simultaneously administer $NaHCO_3$. This is easily done by the Calcium Sandoz Forte which brings 12 mEq of $NaHCO_3$ per tablet of 500 mg of calcium element. In practice we first prescribe calcium Sandoz at increasing doses and once the plasma bicarbonate is normal (25–28 mEq/l) we prescribe additional doses of $CaCO_3$.

Limited Effectiveness of High Doses of $CaCO_3$ in the Case of Hyperparathyroidism and Vitamin D Deficiency Osteomalacia. Two patients of the second study [21] illustrate the limited effectiveness of high doses of $CaCO_3$. One patient had hyperparathyroidism with subperiosteal erosions and plasma calcium between 10.0 and 10.5 mg/dl. High doses of $CaCO_3$ stabilized the X-ray lesions and the levels of alkaline phosphatase but could not achieve complete healing. The other was a 72-year-old woman who developed hip pain while on $CaCO_3$ for 2 years. Her alkaline phosphatases increased and a Looser zone was seen on her pubis. The plasma 25OHD was 4 ng/ml. The administration of 75 μg of $25(OH)D_3$ a week cured the osteomalacia which was actually due to vitamin D deficiency.

These limits led us to systematically supplement dialysis patients with $25(OH)D_3$ at the dose of 75 μg per week given at the first dialysis of the week in order to prevent vitamin D depletion and osteomalacia. As regards the treatment of hyperparathyroidism characterized by subperiosteal resorption and/or high alkaline phosphatases we use small doses of 1α-$(OH)D_3$ and 3 g of $CaCO_3$ eventually associated with $Mg(OH)_2$ as long as these abnormalities are present.

Long-Term Safety of High Doses of $CaCO_3$: Evaluation of the Soft Tissue Calcifications on the Basis of 5 Years Experience. This is our main concern and no data are available in the English reports covering more than a year (only 2 months in Slatopolsky et al. [22]). In our first study vascular calcifications appeared in one case and worsened in another after 6 months: these 2 patients had diabetes and a high plasma calcium phosphate product. In the second study the vascular calcifications worsened in 3 of 12 of the 1α-(OH)D-treated group and in 3 of 15 of the high $CaCO_3$ group. In a recent report [25] vascular calcifications were absent at the discontinuation of $Al(OH)_3$ in 7 patients and remained absent after 36 months of $CaCO_3$ whereas in the 8 patients in whom they were present before high doses of $CaCO_3$, they increased in 4 of them.

Our current experience confirms this finding. We evaluated the calcifications of the aorta and of the iliac and the femoral arteries of 32 patients who had been at least 24 months under control without $Al(OH)_3$ and 1α-$(OH)D_3$ while taking moderate doses of Ca (80 mmol/day as $CaCO_3$ and/or calcium polystyrene sulfonate) associated in 14 of them to $Mg(OH)_2$ at the dose of 2–3 g/day, as well as in a control group treated with $Al(OH)_3$, low doses of Ca (<3 g $CaCO_3$) and if necessary with 1α-$(OH)D_3$. This control group consisted of 30 patients matched for age, sex and previous duration on dialysis. At 24 months as well as at 36 (15 patients) and 48 months (9 patients) there was no significant difference in the increase of the vascular calcifications.

Magnesium Dihydroxyide and Magnesium Carbonate as Phosphate Binders in Patients on Dialysis

$Mg(OH)_2$ and Mg CO_3 are two well-known phosphate binders which have until recently been discarded because of the fear of hypermagnesemia and its immediate hazards on the heart rhythm and its long-term effect on the mineralization process [6]. However, since on a long-term basis high doses of $CaCO_3$ were not sufficient to prevent hyperphosphatemia of 2 mmol/l in about half of our dialyzed patients, Mg phosphate binders became more attractive since other aluminum free phosphate binders are not yet available [27].

We used $Mg(OH)_2$ in 20 patients [28] in whom $CaCO_3$ or Calcium Sorbisterit at the mean daily dose of 100 mmol was insufficient to control the hyperphosphatemia and therefore associated with 1 g ± 1.5 of

$Al(OH)_3$. The latter was discontinued while $Mg(OH)_2$ was introduced and the dialysate magnesium decreased from 0.75 to 0.375 mmol/l in order to prevent hypermagnesemia of ≥ 2 mmol/l. During the 18 months of the study, the mean daily dose of $Mg(OH)_2$ was 2.5 ± 2 g/day while the dose of calcium supplement could be decreased to 75 mmol/day. The plasma magnesium increased moderately to 1.5 ± 0.3 mmol/l whereas the plasma calcium remained constant at 2.43 mmol/l as well as the plasma phosphate at 1.66 ± 0.3 mmol/l. Plasma alkaline phosphatases remained normal and PTH levels remained stable. Furthermore, sequential bone biopsies demonstrated that the slight increase of plasma Mg had not induced a mineralization defect. The main side effect was diarrhea prevented by karaya gum except in one case in which $Mg(OH)_2$ had to be discontinued. Hyperkalemia was a problem in 2 patients requiring $Mg(OH)_2$ discontinuation. This hazard justifies weekly monitoring of plasma magnesium and plasma potassium. In case of hyperkalemia, calcium and not sodium polystyrene sulfonate should be used in order to prevent excessive alcalosis due to the combination of Mg with the resin having free OH^- ions. Our experience is better than that of Guillot [29] who did not associate high $CaCO_3$ and did not decrease the Mg in the dialysate. It is in agreement with that of Donnovan et al. [30] who used magnesium carbonate at a dose of 7.5 ± 2.5 g/day and a magnesium free dialysate so that plasma Mg did not increase. We have been reluctant to use Mg free dialysate fearing Mg depletion in patients treated by $CaCO_3$ alone. Since the decrease of dialysate Mg to 0.375 mmol/l was not sufficient to prevent the slight increase of plasma Mg we would suggest that a dialysate magnesium of 0.25 mmol/l would be optimal since it will keep plasma Mg at a level lower than 1.5 mmol/l in patients taking Mg supplement without inducing hazardous effects in the others [31].

Global Evaluation of a Policy Systematically Eliminating $Al(OH)_3$ in the Prevention of Renal Osteodystrophy in Hemodialyzed Patients on the Basis of 5 Years Experience

Our encouraging experience with moderately high doses of $CaCO_3$ possibly in combination with $Mg(OH)_2$ led us to cancel $Al(OH)_3$ prescription for the past 5 years. Table I summarizes the present clinical and biological status of our whole population currently treated for at least one year, i.e. 57 patients. It shows that $Mg(OH)_2$ was necessary in 26

Table I. Current clinical, biological and therapeutical status of 57 patients treated in Amiens for at least one year

	Duration of dialysis, years (number of patients)					
	1 (12)	2 (11)	3 (9)	4–5 (10)	6–7–8 (6)	10–11–14 (9)
Plasma calcium, mmol/l	2.4	2.4	2.5	2.4	2.3	2.4
Plasma PO_4, mmol/l	1.4	1.8	1.6	1.7	1.5	1.7
Plasma bicarbonate, mmol/l	22	21	22	19	23	26
Plasma Mg, mmol/l	1.1	1.3	1.3	1.4	1.3	1.1
Alk. Pase, IU (N < 170 IU)	121	91	102	97	92	141
Plasma Al, μmol/l	0.4	0.5	0.4	0.3	0.5	0.7
Oral Ca element, mmol/day[a]	34 (11)	42 (8)	41 (7)	72 (10)	97 (5)	52 (6)
Oral $Mg(OH)_2$, g/day	0.5 (3)	1.13 (4)	2.44 (4)	2.65 (6)	2 (4)	1.66 (5)
1α-$(OH)D_3$, μg/week	0 (0)	0.06 (1)	0.14 (2)	0.15 (2)	0.25 (2)	0.2 (2)
Patients with PTX	0	1	0	1	2	3

[a]Number of patients with either $CaCO_3$, $Mg(OH)_2$ or 1α-$(OH)D_3$.

of them inducing only a mild hypermagnesemia (< 1.4 mmol/l). Plasma aluminum was highest in the patients treated for more than 10 years but was only 0.7 μmol/l, i.e. 20 μg/l.

The limit of this treatment to control hyperparathyroidism can be evaluated on the number of patients who necessitated small doses of 1α-$(OH)D_3$ (9 patients) and especially by the number of parathyroidectomies (7 patients). In one case it was a neglected primary hyperparathyroidism; in 2 cases the failure was due to noncompliance so that true failure was actually encountered in 4 cases. In 2 of them severe hyperparathyroidism was already present at the initiation of $CaCO_3$ therapy. Therefore, our approach actually failed in only 2 cases: in a 56-year-old woman on dialysis for 12 years in whom alkaline phosphatase increased in spite of 3 years of $CaCO_3$ therapy; and in a 42-year-old man after 9 years of poor dialysis with major vascular access problems.

The control of hyperparathyroidism could also be evaluated by the follow-up of 11 patients for the 4 last years. Table II shows that PTH levels after a transient decrease rose again to initial levels. The hazard of vascular calcifications has been described above (page 65).

Table II. Long-term follow-up of plasma PTH in 11 patients taking no $Al(OH)_3$ (with yearly data available for 3 years)

Mean ± SEM	Normal	1983	1984	1985	1986
PTH C, pg/ml	20–45	215 ± 42	130 ± 26**	151 ± 28*	187 ± 35
PTH N, pg/ml	80–200	1,820 ± 434	1,320 ± 425	1,245 ± 355***	1,656 ± 392
$CaCO_3$, g/day		9 ± 2	9 ± 2	8.5 ± 2	8.3 ± 2
$Mg(OH)_2$, g/day		2.2 ± 0.5	2.6 ± 0.6	2.7 ± 0.4	2.8 ± 0.4
1α-$(OH)D_3$, 5 μg/day		0	0	2	2

Comparison versus 1983: * $p < 0.05$; ** $p < 0.02$; *** $p < 0.01$.

Calcium Carbonate in the Prevention of Renal Osteodystrophy before Dialysis

In Adults

Although clinically or radiologically evident bone disease is rare before dialysis in adults without vitamin D deficiency, bone biopsy studies have shown that hyperresorption, increased osteoid volume and surfaces and decreased mineralization rate are common. True osteomalacia is however exceptional. Since decreased levels of $1,25(OH)_2D_3$ or 'abnormally normal' levels of $1,25(OH)_2D_3$ may trigger early hyperparathyroidism [32] it seems logical to use small doses of $1,25(OH)_2D_3$ in the prevention of renal osteodystrophy as soon as glomerular filtration rate (GFR) has reached 60 to 40 ml/min. However, early reports have questioned this approach showing that $1,25(OH)_2D_3$ or 1α-$(OH)D_3$ may accelerate deterioration of renal function by inducing hypercalcemia and hyperphosphatemia [33, 34].

Massry et al. [35] have shown, however, that GFR was not depressed after one year of treatment with 0.25–1 μg/day of $1,25(OH)_2D_3$ in 14 patients with mild renal failure in spite of multiple episodes of hypercalcemia and a mean plasma calcium of 10.6 mg/dl. Since bone histology of treated patients improved whereas that of the controls worsened, Massry recommends this preventive approach. Coen proposed a lower dose of $1,25(OH)_2D_3$ (0.25 μg/day); he observed neither hypercalcemia nor hyperphosphatemia and found that the degree of histological hyperparathyroidism was lower than that of a control group with the same degree of renal impairment [36].

Considering that $1,25(OH)_2D_3$ would not only improve calcium absorption but also phosphate absorption and that calcium content of the vessels can increase with 1α-(OH)D therapy at doses not inducing hypercalcemia or hyperphosphatemia [37], we have evaluated another preventive approach, i.e. the use of 3 g of $CaCO_3$ and of nonhypercalcemic doses of $25(OH)D_3$. This treatment was evaluated in 21 uremic patients with an initial creatinine clearance of 41–51 ml/min. The daily dose of $25(OH)D_3$ was 10 μg in 13 patients (group I), 25 μg in 6 patients and 50 μg in 2 patients with a plasma calcium < 2.25 mmol/l (group II). $Al(OH)_3$ was not associated but restriction of diary products was advised. Plasma 25(OH)D increased from 12 to 24 ng/ml in group I and from 15 to 45 ng/ml in group II. Plasma creatinine increased mildly from 3.5 ± 1.0 to 5.0 ± 4.5 mg/dl the second year. The slope of plasma creatinine, however, decreased from 50 to $30 \cdot 10^{-3}$ mg/dl/month after initiation of treatment. Plasma PTH (C terminal) increased not significantly from 140 ± 70 to 180 ± 70 pg/ml the second year. Alkaline phosphatase decreased from 150 ± 70 to 130 ± 40 IU ($p < 0.05$). Plasma calcium remained stable at 2.35 ± 0.1 mmol/l throughout the study in both groups. Plasma PO_4 increased from 1.1 to 1.5 mmol/l ($p < 0.05$). Urinary calcium increased from 1 to 2 mmol/day. Plasma bicarbonate remained constant at 25 ± 1 mmol/l. A bone biopsy after double tetracycline labeling was performed after 15 ± 3 months of prevention in 12 patients. Osteoid volume decreased from high values (5 ± 3) to normal value as well as osteoid surface. Osteoid thickness and mineral apposition rate remained normal. Active resorption surface and osteoblast surface decreased. Metastatic calcification occurred in only one patient with existing ischemic heart disease.

Thus, this study shows that in adult uremic patients before dialysis, 3 g of $CaCO_3$ and nonhypercalcemic doses of $25(OH)D_3$, can prevent osteomalacia and acidosis and suppress, although not completely, hyperparathyroidism without accelerating the decline of renal function. The histological results may appear less impressive than those of Massry but were achieved with no frequent hypercalcemia which is safer as regards the vascular calcifications. Compared to the data of Coen, the biochemical parameters were quite comparable as regards plasma calcium, plasma PO_4, calciuria, $1,25(OH)_2D_3$ and the decrease in alkaline phosphatases; the histological ones were not as good as regards bone resorption but were better as regards the extent of osteoid volume and osteoid surface.

In uremic patients in whom vitamin D deficiency is prevented, a controlled trial comparing low doses of $1\alpha\text{-(OH)}D_3$ or of $1,25(OH)_2D_3$ to $CaCO_3$ at doses inducing the same levels of calcium in the plasma and the urine is necessary.

$CaCO_3$ as Phosphate Binder in Children before Dialysis

Children are particularly susceptible to aluminum toxicity from aluminum phosphate binders [6]. Therefore, Alon et al. [39] tried to treat renal osteodystrophy of children not yet on dialysis (serum creatinine between 3.7 and 5.1 mg/dl by high doses of $CaCO_3$ (0.1–0.3 mg/kg BW) in association with small doses of $1\alpha\text{-(OH)}D_3$ plasma calcium increased from 8.9 to 10.3 mg/dl and plasma phosphate decreased from 6.3 to 4.2 mg/dl. Plasma alkaline phosphatase decreased and bone X-rays showed either healing of renal osteodystrophy or its prevention. No change in serum creatinine concentrations was noted and only one episode of mild hypercalcemia occurred. They concluded therefore that $CaCO_3$ is an effective and safe substitute for $Al(OH)_3$ resulting in an improved bone and mineral homeostasis.

Conclusions

Calcium carbonate at high doses (3–14 g/day) taken with the meals is an efficient and safe phosphate binder provided plasma levels of calcium and phosphate are monitored at 1–2 weeks intervals in dialysis or at 1–3 months interval before dialysis. It allows prevention of hyperparathyroidism worsening in most adult patients without $1\alpha\text{-(OH)}D_3$. In children needs for vitamin D are greater and a dose of 0.25 μg/day of $1\alpha\text{-(OH)}D_3$ or $1,25(OH)_2D_3$ is recommended. In patients on dialysis, high doses of $CaCO_3$ alone may not be sufficient even when dietary phosphate restriction is reinforced and dialysis prolonged. In these cases $Mg(OH)_2$ or magnesium carbonate may be added while dialysate magnesium is decreased to 0.25 mmol/l or lower to prevent hypermagnesemia. In order to induce a greater tolerance to high doses of $CaCO_3$, dialysate calcium and calcium in the substitution fluid for hemofiltration should be adapted to obtain a zero calcium balance. We recommend 6.0 mg/dl for dialysate calcium since the normal ultrafiltered calcium is 6.0 mg/dl. In patients previously loaded with aluminum, desferrioxamine treatment may improve the tolerance to high doses of $CaCO_3$ since aluminum osteopathy is a cause of hypercalcemia. Following these recommendations an aluminum free control of osteodystrophy can be achieved without significant increase in soft tissue calcification after 4 years.

References

1 Fournier, A.; Morinière, P.; Sebert, J.L.; Boudailliez, B.; Gregoire, I.; Garabedian, M.; Guéris, J.; Meunier, P.: Renal osteodystrophy: general concepts and current issues; in Nephrology, vol. II, pp. 1357–1373 (Robinson, Springer, New York 1985).

2 Popovtzer, M.M.; Michael, U.F.; Joanson-Dial, K.S.; Ronstadt, C.; Nelson, D.; Ogden, D.A.: Dietary calcium deprivation and secondary hyperparathyroidism in patients treated by chronic dialysis. Mineral Electrolyte Metab. *12:* 298–302 (1986).

3 Slatopolsky, E.; Lopez-Hilker, S.; Dusso, A.; Morrissey, J.J.; Martin, K.J.: Secretion of parathyroid hormone in chronic renal failure; in De Broe, Van de Vyver, Bone and renal failure. Contr. Nephrol., vol. 64, pp. 16–24 (Karger, Basel 1988).

4 Bordier, P.; Tunchot, S.; Eastwood, J.B.; Fournier, A.; Wardener, H.E. de: Lack of histological evidence of vitamin D abnormality in the bones of anephric patients. Clin. Sci. *44:* 33–41 (1973).

5 Ward, M.K.; Ellis, M.A.; Feest, T.G.; Kerr, D.: Osteomalacia dialysis osteodystrophy. Evidence for a waterborne aetiological factor, probably aluminium. Lancet *i:* 841–845 (1978).

6 Fournier, A.; Morinière, P.; Sebert, J.L.; Dkhissi, H.; Atik, A.; Leflon, P.; Renaud, H.; Guéris, J.; Gregoire, I.; Idrissi, A.; Garabedian, M.: Calcium carbonate, an aluminium free agent for control of hyperphosphatemia, hypocalcemia and hyperparathyroidism in uremia. Kidney int. *29:* suppl. 18, pp. S114–S119 (1986).

7 Memmos, D.E.; Eastwood, J.B.; Harris, E.; O'Grady, A.; Wardener, H.E. de: Response of uremic osteoid to vitamin D. Kidney int. suppl. 21, pp. S50–S54 (1982).

8 Morapalma, F.I.; Ellis, H.A.; Cook, D.B.; Dewar, J.H.; Ward, E.K.; Wilkinson, R.; Kerr, D.N.S.: Osteomalacia in patients on chronic renal failure before dialysis or transplantation. Q. Jl Med. *52:* 332–348 (1983).

9 Nielsen, H.E.; Melsen, F.; Christensen, M.S.: Interrelationship between calcium phosphorus metabolism, serum parathyroid hormone and bone histomorphometry in nondialyzed and dialyzed patients with chronic renal failure. Mineral Electrolyte Metab. *4:* 113–122 (1980).

10 Tahiri, Y.; Morinière, P.; Jaudon, M.C.; Dkhissi, H.; Fournier, A.: Hyperaluminémie des hemodialysés chroniques. Evaluation du rôle respectif de l'aluminium du dialysat et de celui de la prise orale de gel d'alumine. Nephrologie *4:* 129–133 (1983).

11 Sebert, J.L.; Fournier, A.; Fohrer, P.; Fremont, J.F., de; Morinière, P.: Comparative evaluation of bone aluminium content and of bone histology in patients on chronic hemodialysis and chronic hemofiltration. Nephron *42:* 34–40 (1986).

12 Fournier, A.; Arnaud, C.D.; Johnson, W.J.; Goldsmith, R.S.: Etiology of hyperparathyroidism and bone disease during chronic hemodialysis. J. clin. Invest. *50:* 599–605 (1971).
13 Clarkson, E.M.; McDonald, S.J.; Wardener, H.E. de: The effects of high intake of calcium carbonate in normal subjects and patients with chronic renal failure. Clin. Sci. *30:* 425–438 (1966).
14 Curtis, J.R.; Wardener, H.E. de; Gower, P.E.; Eastwood, B.: The use of calcium carbonate and phosphate without vitamin D in the management of renal osteodystrophy. Proc. EDTA *1970:* 141–145.
15 Meyrier, A.; Marsac, J.; Richet, G.: The influence of a high calcium carbonate intake on bone disease in patients undergoing hemodialysis. Kidney int. *4:* 146–153 (1973).
16 Roussel, A.: Evaluation à long terme de l'ostéodystrophie des hémodialysés chroniques sans supplément vitaminique D; thèse de doctorat de Médecine, Amiens (1982).
17 Renaud, H.; Atik, A.; Benelmoufok, S.; Fournier, A.: Evaluation of vascular calcinosis risk in patients on chronic hemodialysis. Lack of influence of calcium carbonate doses. Nephron *48:* 28–32 (1988).
18 Makoff, D.L.; Gordon, A.; Franklin, S.; Gerstein, A.; Maxwell, M.H.: Chronic calcium carbonate therapy in uremia. Archs intern. Med. *123:* 15 (1969).
19 Papadoyannakis, N.J.; Stefanidis, C.J.; Patricarea, A.: Kotsovaslis, K.; Georgiopoulou, A.: The effect of calcium carbonate administration in nitrogen metabolism in patients on haemodialysis. Proc. EDTA *22:* 83–87 (1985).
20 Morinière, P.; Roussel, A.; Tahiri, Y.; Fournier, A.: Substitution of aluminium hydroxide by high doses of $CaCO_3$ in patients on chronic hemodialysis: disappearance of hyperaluminemia and equal control of hyperparathyroidism. Proc. EDTA *19:* 784–787 (1982).
21 Morinière, P.; Fournier, A.; Leflon, A.; Sebert, J.L.; Gregoire, I.; Bataille, P.; Guéris, J.: Comparison of 1α OH vitamin D_3 and high doses of calcium carbonate for the control of hyperparathyroidism and hyperaluminemia in patients on maintenance dialysis. Nephron *39:* 309–315 (1985).
22 Slatopolsky, E.; Weerts, C.; Lopez-Hilker, S.; Norwood, K.; Zink, M.; Windus, D.; Delmez, J.: Calcium carbonate as a phosphate binder in patients with chronic renal failure undergoing dialysis. New Engl. J. Med. *315:* 157–161 (1986).
23 Salusky, I.B.; Coburn, J.W.; Foley, J.; Nelson, P.; Fine, R.N.: Calcium carbonate as a phosphate binder in children on dialysis: effect on plasma aluminium. J. Pediat. *108:* 767–780 (1986).
24 Hercz, G.; Kraut, J.; Andress, D.A.; Howard, N.; Roberts, C.; Shinaberger, J.H.; Sherrard, D.J.; Coburn, J.W.: Use of calcium carbonate as a phosphate binder in dialysis patient. Mineral Electrolyte Metab. *12:* 314–319 (1986).
25 Monnier, N.; Bournerias, F.; Reveillaud, R.J.: Hémodialyse à long-terme sans aluminium; in Traeger, Survie à long terme des hémodialysés, pp. 155–158 (Gambro, Paris 1986).
26 Ramirez, J.A.; Emmet, H.; White, M.G.; Fathi, N.; Santa Ana, C.A.; Moranski, S.G.; Fordtran, J.S.: The absorption of dietary phosphorus and calcium in hemodialysis patients. Kidney int. *30:* 753–759 (1986).

27 Schneider, H.W.; Kulbe, K.D.; Weber, H.; Streicher, E.: In vitro and in vivo studies with a non-aluminium phosphate binding compound. Kidney int. *28:* S120–S12 (1986).

28 Morinière, P.; Vinatier, I.; Colome, M.F.; Abdulmassih, Z.; Atik, A.; Hocine, C.; Marie, A.; Leflon, P.; Guéris, J.; Fournier, A.: Control of hyperphosphatemia in chronic haemodialysis by calcium carbonate and magnesium hydroxide. Prevention of hyperaluminemia without bone toxicity. Nephrology Dialysis Transplant. (in press, 1988).

29 Guillot, A.P.; Hood, V.L.; Rungel, P.; Gennari, S.: The use of magnesium containing phosphate binders in patients with end stage renal disease on maintenance dialysis. Nephron *30:* 114–117 (1982).

30 O'Donnovan, R.; Monitz, C.; Baldwin, D.: Control of hyperphosphatemia by oral magnesium carbonate on zero magnesium dialyzate without aluminium binders. Lancet *i:* 880–881 (1986).

31 Gonella, M.: Plasma and tissue levels of magnesium in chronically hemodialyzed patients. Effect of dialysate magnesium levels. Nephron *34:* 141–145 (1983).

32 Llach, F.; Massry, S.G.: On the mechanism of secondary hyperparathyroidism in moderate renal insufficiency. J. clin. Endocr. Metab. *61:* 601–606 (1985).

33 Christiansen, C.; Rodbro, P.; Christensen, M.S.: Deterioration of renal function during treatment of chronic renal failure with $1,25(OH)_2$ vitamin D_3. Lancet *ii:* 700–703 (1978).

34 Toogard, L.; Sorensen, E.; Brochner Mortensen, J.: Controlled trial of 1-alpha-hydroxycholecalciferol in chronic renal failure. Lancet *i:* 1044–1047 (1976).

35 Massry, S.G.: Assessment of $1,25(OH)_2D_3$ in the correction and prevention of renal osteodystrophy in patients with mild to moderate renal failure; in Schaefer, Vitamin D, chemical, biochemical and clinical update, pp. 935–937 (de Gruyter, Berlin 1985).

36 Coen, G.; Mazzaferro, S.; Bonucci, E.; Ballant, I.D.: Treatment of secondary hyperparathyroidism of predialysis chronic renal failure with low doses of $1,25(OH)_2D_3$: humoral and histomorphometric results. Mineral Electrolyte Metab. *12:* 375–382 (1986).

37 Tvedegard, E.; Ladefoged, O.; Nielsen, M.; Kramstrup, O.: Effect of 1α OH vitamin D_3 and dietary calcium and phosphate on the aortic mineral content in rabbits with mild azotemia. Nephron *34:* 185–191 (1983).

38 Fournier, A.; Idrissi, A.; Sebert, J.L.; Marie, A.; Guéris, J.; Garabedian, M.: Prevention of osteodystrophy by calcium carbonate and non hypercalcemic doses of 25 OH vitamin D_3 in patient with moderate renal failure. Kidney int. (in press, 1988).

39 Alon, U.; Davidai, G.; Bentur, L.; Berant, M.; Better, O.: Oral calcium carbonate as phosphate binder in infants and children with chronic renal failure. Mineral Electrolyte Metab. *12:* 320–325 (1986).

A. Fournier, MD, Service de Néphrologie-Médecine interne et Ranimations,
CHU d'Amiens, F-80030 Amiens (France)

Contr. Nephrol., vol. 64, pp. 74–82 (Karger, Basel 1988)

The Cellular Metabolism of Lead and Calcium: A Kinetic Analysis in Cultured Osteoclastic Bone Cells

John F. Rosen [a], *Joel G. Pounds* [b]

[a]Department of Pediatrics, Albert Einstein College of Medicine, Montefiore Medical Center, New York City, N.Y.; [b]Department of Applied Science, Brookhaven National Laboratory, Upton, N.Y., USA

Bone is the major long and short-term reservoir for exchangeable lead. Concentrations of lead in bone correlate closely with the presence of lead nephropathy in adults [1, 2] and impairments in neuropsychological development in children [3]. Moreover, a recent analysis of the NHANES II data set in children indicated an inverse correlation between blood lead (range: 5–35 μg/dl) and stature; for each 10 μg/dl increment in concentration, a 1.2-cm decrease in height was observed [4]. Moreover, serum concentrations of osteocalcin, a calcium-binding protein unique to bone and synthesized by osteoblasts, were markedly elevated in lead intoxicated children (blood lead: 25–55 μg/dl), as compared to age-matched controls, $p<0.001$ [5]. We postulated that lead may inhibit osteocalcin binding to bone, thereby allowing osteocalcin to enter the plasma at relatively high concentrations. Indeed, we found that lead (5 μ*M* or higher) inhibited the binding of osteocalcin to bone hydroxy-apatite in vitro [5]. Furthermore, we have recently observed that lead displaces calcium from osteocalcin with a greater affinity for this protein compared to that of calcium [5]. These data are similar to lead's higher affinity for other calcium-binding proteins, including calmodulin, compared to that of calcium [6]. Taken together, these observations indicate that the skeleton should be considered a target tissue for this toxic metal, as well as an important toxicological source of lead.

Calcium is recognized as an essential intracellular messenger in the stimulus response-cascades of several hormone target-cell systems [7]. In addition, calcium acts symmetrically as a modulator of cyclic nucleotide metabolism thereby participating with adenosine 3′:5′-cyclic phosphate (cAMP) in effecting cellular responses to several hormones [7]. Furthermore, there is mounting evidence that calcium may mediate some of the skeletal responses to calciotropic hormones [7, 8]. To date, a detailed study of calcium metabolism in isolated bone cells has yet to be reported; and studies to characterize bone cell calcium homeostasis have yielded conflicting results [8–10]. The present studies were designed to characterize the steady-state kinetic behavior of lead and calcium in bone cell populations enriched for osteoclastic bone cells. By so doing, cellular constituents associated with the kinetic compartments could be identified; and the cellular metabolism of these two metals in osteoclastic bone cells could be compared to cells derived from a soft tissue (hepatocytes) [11–14].

Bone cells, derived from mouse calvaria, were enriched for osteoclatic bone cells by a sequential collagenase digestion [13–15] and maintained in primary culture for one week. Osteoclastic bone cells, in separate experiments from those to characterize ^{45}Ca washout curves and kinetic parameters, were labeled with 5 μ*M* lead (^{210}Pb–1.0 μCi/ml) for 20 h, subsequently washed and medium without radiotracer was added to each well. At frequent time intervals to 210 min, medium was removed for scintillation counting and 500 μl of fresh medium was added at each time point. (1) *Pool sizes in nmol · mg protein*$^{-1}$ *(S_1)*, (2) *fluxes between compartment i and j in nmol · min*$^{-1}$ *mg · protein*$^{-1}$ *(Pij)*, (3) *rate constants in min*$^{-1}$ *(kij)*, and (4) *half-times in min (Tij)* were determined for ^{210}Pb using a three-compartment intracellular model [13]. Pool sizes and transfer functions, normalized to 1.0 mg cell protein, were derived from the coefficients and exponents for a three-term sum of exponential equation for individual cultures within an experiment, or expressed as the mean ± SEM for replicate experiments [11–14]. The parameters derived from treated cultures were compared to those obtained from untreated control cultures by the multiple comparison method of Dunnett. A detailed rationale for the use of the three compartment model is presented elsewhere [11–14] (fig. 1).

Using these techniques, we described a three-compartment model for the steady-state metabolism of lead in primary cultures of osteoclastic bone cells [13]. This compartmental analysis provides information regarding the number of kinetic compartments and the transfer functions associated with these compartments. Little information is provided regarding the specific relationship of the biological or biochemical constituents associated with the kinetic compartments; and each kinetic pool may not coincide with a specific structure (organelle) or biochemical entity. Fortunately, many metals, including lead and calcium, are relatively specific for certain organelles, storage proteins or transport pathways, which can be manipulated by hormones and various agents. Based

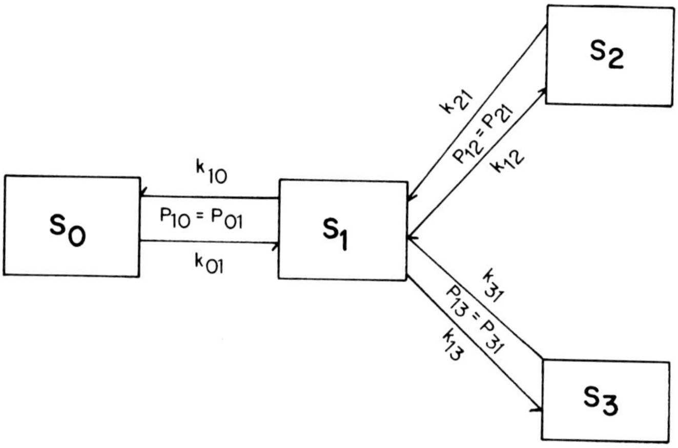

Fig. 1. Three compartmental model for the steady-state cellular metabolism of lead and calcium. S_0 represents Pb^{2+} or Ca^{2+} in the cell culture medium and S_1, S_2 and S_3 represent three intracellular kinetic pools of exchangeable Pb^{2+} or Ca^{2+}.

upon data from several desaturation experiments, the intracellular metabolism of lead in osteoclastic bone cells was summarized into a physiological model (fig. 2) [13]. Though the exact anatomical characterization of each compartment is incomplete, this model is relevant to the design of other studies concerning the regulation of cellular lead metabolism by calciotropic hormones and intracellular messengers.

Changing medium phosphate (P_i) concentrations had biphasic effects on the kinetic distribution of cellular lead. All three kinetic pools of cellular lead increased in size in medium containing 1–6 m*M* P_i but higher P_i concentrations effected a decrease in S_2, S_3 and S_{Total}. These findings are best explained by perturbations in cellular calcium homeostasis at elevated medium P_i levels. Incubation of osteoclastic bone cells in 8 m*M* P_i markedly increased cell calcium (1000%), as determined by similar kinetic methodology. This very large increase in cellular calcium may permit effective competitive inhibition of lead transport into mitochondria by calcium, as noted in mitochondria from other tissues.

The three kinetic pools of lead were defined as intracellular by their refractoriness to chelation by ethyleneglycoltetraacetic acid (EGTA) and $LaCl_3$ [13]. Accordingly, K_{10}, T_{10}, P_{10} and P_{01} (fig. 1) represent the transfer functions for lead movement across the plasma membrane. The relatively

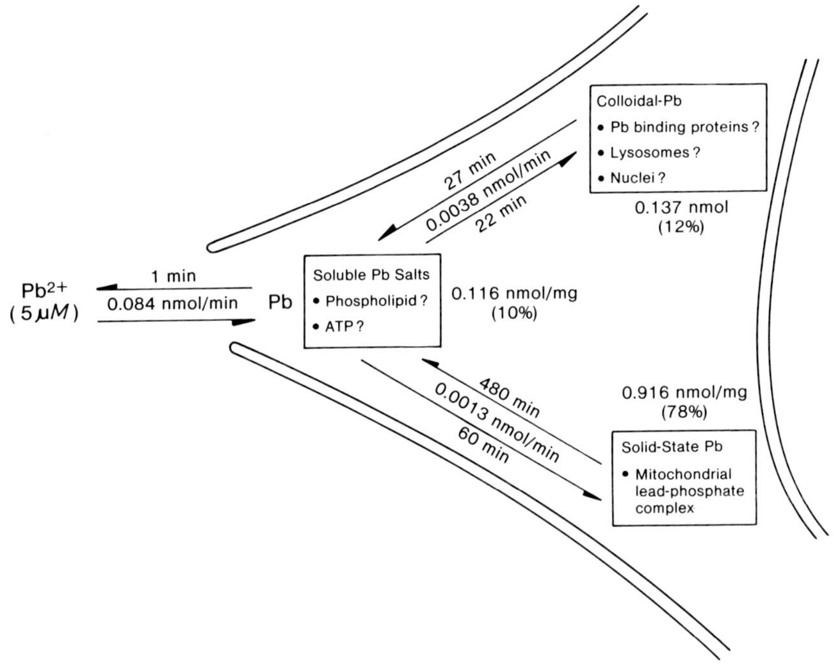

Fig. 2. Physiological model for the steady-state intracellular metabolism of lead in osteoclastic bone cells. From ref. [13].

long half-time of T_{31} and the high capacity and the sensitivity to KCN treatment of S_3 indicated that this kinetic compartment includes mitochondria lead (fig. 2). K_{13}, K_{31}, T_{13}, T_{31}, P_{13} and P_{31} represent transfer functions for lead movement into and out of mitochondria. This detailed understanding of lead metabolism in primary cultures of osteoclastic bone cells provides a new and unique biological framework for assessing the regulation of bone lead in vivo.

Experiments were also carried out to evaluate the effect of lead concentration in the medium on cellular lead homeostasis. Lead was accumulated by osteoclastic bone cells in a concentration-dependent manner between 1 and 50 μ*M* lead [13]. The concentration-dependent nature of intracellular lead metabolism is shown by similar changes in rate constants [13]. The rate constants describing lead exchange between S_1 and S_2 did not change, whereas K_{10} increased, as K_{31} and K_{13} decreased with increasing lead concentration. Lead associated with the slowly exchanging pool (S_3) markedly increased, although none of the kinetic pools became saturated [13].

These data further delineate the dynamic nature of lead metabolism in bone, specifically in osteoclastic bone cells. The results indicated that exchange of lead across the plasma membrane of osteoclasts readily occurs, the largest of these compartments (S_3) was tentatively identified as including mitochondrial lead and the intracellular distribution of lead in osteoclastic bone cells was similar to that in hepatocytes [11–14]. In fact, the overall qualitative similarities were remarkable between osteoclastic bone cells and hepatocytes. In both cell types, 75% or more of cellular lead was in S_3, with the remainder divided almost equally between the more rapidly exchanging compartments, S_1 and S_2. Of the five rate constants defining the steady-state exchange, only K_{31} and T_{31} were statistically different between the two cell types. These similarities suggest that differences in sensitivity of target organs to the toxic effects of lead may depend more on the specific metabolic role(s) and function of specific cell types than on the distribution and/or metabolism of lead within different cells [1, 2, 11–13].

These studies provide considerable support for the concept that lead entry and exit into and out of bone occurs in a dynamic manner similar to that of calcium. These studies also provided a framework to examine the cellular metabolism of calcium in osteoclastic bone cells, the effects of lead on the distribution of calcium within osteoclasts and possible hormonal modulators of lead metabolism in these cells.

Before the specific role(s) of calcium as an intracellular messenger in carrying out the cellular actions of calciotropic hormones can be delineated and before lead-induced perturbations in cellular calcium homeostasis could be understood in resting and hormonally stimulated osteoclastic bone cells, studies were conducted to characterize the steady-state kinetic behavior of calcium in primary cultures of these cells [14]. Compartmental modeling of calcium was carried out as previously described for lead [11–13], except that the specific activity of the compartments was normalized for different labeling times. To our knowledge, our results are the very first to resolve three kinetic pools of intracellular calcium in bone cells (in general) and in osteoclastic bone cells (specifically). Previous reports using fibroblasts, mixtures of bone cell populations, osteosarcoma cell lines or kidney cells delineated only two intracellular kinetic pools. Data from our experiments were best fit by a three-term sum of exponential equations. Even when surface bound calcium was chelated by EGTA, two exponential terms were inadequate to fit the data [14]; and increasing the number of terms from three to four did not improve the fit. This resolution of three kinetic pools of intracellular calcium was attributed to shorter time intervals early in the washout-desaturation procedure and the use of higher specific activity ^{45}Ca.

A brief rinse with 0.5 m*M* EGTA immediately before the washout removed calcium bound to osteoclastic bone cell surfaces; and this rinse was included in all subsequent experiments. Hence, all of the kinetic compartments were intracellular; and K_{10}, T_{20} and P_{10} defined calcium exchange across the plasma membrane [14]. Ruthenium red, an inhibitor of calcium transport into mitochondria, reduced the size of the slowest exchanging pool (S_3) by about 40% and altered the transfer functions defining the exchange of calcium between S_3 and S_2, thereby indicating that S_3 included mitochondrial calcium. This conclusion was also supported by the finding that increasing the concentration of P_i in the medium markedly increased the size of S_3, probably as a result of calcium-P_i deposition in the mitochondrial matrix.

We have postulated that perturbations in cellular calcium homeostasis produced by lead at relatively low concentrations may be an early and discrete toxic effect of lead common to diverse cell types [1, 2, 11–13]. The steady-state kinetic distribution of calcium in osteoclastic bone cells as a function of medium lead concentrations has been examined [Pounds and Rosen, in preparation]. Rising medium concentrations of lead (0–50 µ*M*) produced a marked increase in total cellular calcium (8.094–46.789 nmol/mg cell protein). Increments in all three intracellular calcium compartments occurred in a dose-response manner, as medium lead levels rose; and the largest increase was observed in S_3, previously delineated as mitochondrial calcium. These data indicate that lead perturbs calcium homeostasis in osteoclastic bone cells. As a result, the regulation of calcium-dependent cellular functions may be placed out of the physiological range of normal control through changes in intracellular ionic calcium concentrations.

Direct clinical correlates of these in vitro data may ultimately be provided by measuring cytosolic calcium and lead concurrently in readily available cells (lymphocytes, platelets) from lead-intoxicated humans by ^{19}F NMR. In the interim, the toxicological importance of lead in bone is currently being measured by L-X-ray fluorescence (XRF) measurements [16, 17] in the tibia of lead-intoxicated children [18]; and these measurements are being assessed in relation to biochemical, neurobehavioral and electrophysiological indices of lead toxicity.

The first description of an XRF system designed to measure soft L-X-rays of lead, in which such measurements were compared by electrothermal atomic absorption spectroscopy (EAAS), was reported by Wielopolski et al. [17]. The relationship between the net number of counts in the lead peaks of the tibia, measured with an ^{125}I source, and the concentrations of lead in bone, determined by EAAS, yielded a correlation coefficient of 0.90 ($p < 0.03$); and the detection limit for lead was 22 pp or three times the standard deviation of the background [17]. To further validate these XRF measurements, the solubilized bone samples (used for EAAS) were reanalyzed by XRF and a similar r value was found. The fluctuations due to the overlying tissue were accounted for with a real-time ultrasound unit (7 MHz). Since cortical bone thickness is more than the saturation thickness of 3.0 mm, the signal is independent of bone density; and a constant volume of bone cortex was sampled by each measurement. Wielopolski et al. [17] pointed out and proved experimentally that a polarization technique (using a low energy X-ray tube, for instance) could be utilized to reduce the background and thereby improve the signal-to-noise ratio. Furthermore, the very low radiation exposure from a system using soft L-X-rays from a low energy X-ray tube suggested the possibility of its use in children, an age-group for which institutional review board approval in the USA is difficult to obtain for any radiation exposure that is not immediately diagnostic and/or therapeutic.

Very recently, this L-X-ray instrument has been redesigned; and a low-energy X-ray source is now used to irradiate the tibia [18]. The beam has been polarized; and the signal-to-noise ratio has been increased severalfold with careful attention to geometry. The new detection limits are less than 5 ppm with a skin dose of about 1.0 rad and virtually no radiation dose to the bone marrow. These marked improvements in L-X-ray detection limits suggest that it may now be possible to incorporate L-XRF measurements sequentially in lead-intoxicated children into a comprehensive multidisciplinary clincical research protocol.

Acknowledgements

Parts of this article are based upon collaborative efforts with several investigators. These include Drs. M. Markowitz, H. Ruff, D. Slatkin, L. Wielopolski, H. Cohen, H. Vaughan, D. Kurtzberg, and P. Bijur. Preparation of this manuscript was supported by NIH Grants ES 01060, ES 04039, ES 04040, Maternal Child Health Grant MCJ 360488 and DOE Contract DE-AC02-76CH00016. By acceptance of this article, the publisher and/or recipient acknowledges the US Government's right to retain a nonexclusive, a royalty-free license in and to any copyright covering this paper.

References

1 Rosen, J.F.: Metabolic and cellular effects of lead: a guide to low level lead toxicity in children; in Mahaffey, Dietary and environmental lead exposure, pp. 157–185 (Elsevier, Amsterdam 1985).

2 Rosen, J.F.: The toxicological importance of lead in bone: the evaluation and potential uses of bone lead measurements by X-ray fluorescence to evaluate treatment outcomes in modeately lead toxic children; in Clarkson, Scientific basis and practical applications of biological monitoring of toxic metals (Plenum Press, New York 1986).

3 Needleman, H.L. (ed.): Low level lead toxicity (Raven Press, New York 1980).

4 Schwartz, J.; Angle, C.R.; Pirkle, J.L.; Pitcher, H.: Childhood blood lead levels and stature. Pediatrics *77:* 281–288 (1986).

5 Markowitz, M.E.; Gundberg, C.M.; Rosen, J.F.: Sequential osteocalcin sampling as a biochemical markers of the success of treatment in moderately lead poisoned children. Pediat. Res. (in press, 1988).

6 Fullmer, C.S.; Edelstein, S.; Wasserman, R.H.: Lead-binding properties of intestinal calcium-binding proteins. J. biol. Chem. *260:* 6816–6819 (1985).

7 Rasmussen, H.: The calcium messenger system. New England Journal of Medicine *314:* 1094–1101 (1986).

8 Dziak, R.; Stern, P.H.: Calcium transport in isolated bone cells: Effects of parathyroid hormone and cyclic 3′, 5′ AMP. Endocrinology *97:* 1281 (1975).

9 Lowik, C.W.G.M.; Leeuwen, J.P.T.M. van; Meer, J.M. van der; Zeeland, J.K. van; Scheven, B.A.A.; Herrmann-Erlee, M.P.M.: A two receptor model for the action of parathyroid hormone on osteoblasts: A role for intracellular free calcium and cAMP. Cell Calcium *6:* 311 (1985).

10 Eilam, Y.; Szydel, N.; Harell, A.: Effects of vitamin D metabolites on cellular Ca^{2+} and on Ca transport in primary cultures of bone cells. Mol. cell. Endocrinol. *18:* 215 (1980).

11 Pounds, J.G.; Wright, R.; Kodell, R.L.: Cell metabolism of lead: A kinetic analysis in the isolated rat hepatocyte. Toxicol. appl. Pharmacol. *66:* 88–101 (1982).

12 Pounds, J.G.: Effect of lead intoxication on calcium homeostasis and calcium-mediated cell function. Neurotoxicology *5:* 295–332 (1984).

13 Pounds, J.G.; Rosen, J.F.: Cellular metabolism of lead: A kinetic analysis in cultured osteoclastic bone cells. Toxicol. appl. Pharmacol. *83:* 531–545 (1986).

14 Rosen, J.F.; Pounds, J.G.: Quantitative interactions between lead and calcium in cultured osteoplastic bone cells. Toxicol. Appl. Pharmacol. (in press, 1988).

15 Wong, G.L.; Kent, G.N.; Ku, K.Y.; Cohn, D.V.: The interaction of parathyroid hormone and calcium on the hormone-regulated synthesis of hyaluronic acid and citrate decarboxylation in isolated bonc cells. Endocrinology *103:* 2274 (1978).

16 Wielopolski, L.; Rosen, J.F.; Slatkin, D.N.; Vartsky, D.; Ellis, K.J.; Cohn, S.H.: Feasibility of non-invasive analysis of lead of the human tibia by soft L-x-ray fluorescence. Med. Phys. *10:* 248–251 (1983).

17 Wielopolski, L.; Slatkin, D.N.; Vartsky, D.; Ellis, K.J.; Cohn, S.H.: Feasibility study for the in vivo measurement of lead in bone using L-x-ray fluorescence. IEEE Trans. Nuc. Sci. *28:* 114–116 (1981).

18 Rosen, J.F.; Jenks, S.T.; Markowitz, M.E.; Slatkin, D.N.; Wielopolski, L.: L-x-ray fluorescence: A rapid assessment of cortical bone lead in lead-toxic children. Toxicologist *8:* 25 (1988).

John F. Rosen, MD, Department of Pediatrics, Albert Einstein College of Medicine, Montefiore Medical Center, 111 E. 210th Street, Bronx, New York City, NY 10467 (USA)

Contr. Nephrol., vol. 64, pp. 83–92 (Karger, Basel 1988)

Mechanisms of Metal-Induced Renal Cell Injury: Roles of High-Affinity Metal-Binding Proteins

Bruce A. Fowler

National Institute of Environmental Health Sciences,
Research Triangle Park, N.C., USA

The mechanisms of metal-induced renal cell injury are highly complex. At present, determination of the biologically active chemical forms of these agents is a major challenge in elucidating toxic manifestations of metals. This information is also obviously critical to estimating the target dose of a metal required to produce toxicity at the molecular, cellular and tissue levels. For common toxic metals such as Cd and Pb, extensive intracellular binding to high-affinity molecules such as metallothionein (MT) and the more recently discovered lead-binding proteins (PbBP) within kidney tubule cells is of central importance towards understanding the mechanisms of toxicity and interactions between toxic/essential metals. Recent studies also suggest that binding of metals to high affinity proteins may also play a critical role in facilitating the movement of metals across organelle membranes within cells thereby permitting these agents to interact with a greater number of essential molecular systems.

The present review will attempt to briefly examine the known and hypothesized roles of MT and PbBP in mediating nephrotoxicity from these two metals alone or in combination since multielement exposure are a common occurrence under both occupational and environmental circumstances. Possible directions for future research in the area of metal-induced nephrotoxicity will also be considered.

Metallothionein

The low molecular weight metal-binding protein, MT has been the subject of extensive study for 30 years [1] and although it has been chemically well-characterized (table I), the normal physiological roles of this molecule within cells have yet to be elucidated. Studies from a number of laboratories [2–5] have suggested , however, that MT plays a major role in regulating the bioavailability of essential divalent metal cations such as Zn^{2+} [2–4] and Cu^{2+} [5] to metalloenzymes requiring these metals. The possible significance of this postulated function in mediating the nephrotoxic potential of toxic metal ions such as Cd^{2+} and Pb^{2+} is discussed below in relation to PbBP.

During recent years, MT has been shown to play several roles in the nephrotoxicity of Cd^{2+}. The development of radioimmunoassays for MT [6, 7] has demonstrated its presence in the circulation of both humans and animals exposed to Cd^{2+}. Since MT is a low molecular weight protein which is readily filtered by the glomerulus and efficiently reabsorbed by the proximal tubule cells, these findings delineated the mechanism by which Cd^{2+} was delivered to the target cells of the kidney. Other studies [8–16] have shown that the reabsorbed CdMT complex is handled by the proximal tubule cells in a manner similar to other low molecular weight proteins [17] with rapid degradation of the complex resulting in the release of Cd^{2+} ions into the cytosol. This process is followed by the de novo synthesis of renal MT with subsequent re-accumulation of the Cd^{2+} in this compartment.

The nephrotoxicity of chronic Cd^{2+} exposure in humans and animals is primarily characterized by the development of low molecular weight tubular proteinuria and calciuria [16, 18–22]. In some cases, osteomalacia is also observed [18–20] suggesting that perhaps the renal calciuria may play a role in this process. The mechanism(s) underlying these phenomena and the role of MT have been difficult to study due to the long time period required for development of tubule lesions. Recently, a parenteral CdMT injection model [9, 11–14, 16, 21–25] which takes advantage of the rapid delivery of CdMT to proximal tubule cells has been developed to study the mechanisms of Cd^{2+} nephrotoxicity within a short time frame. The ultrastructural effects of CdMT administration on the renal proximal tubule cells are characterized by the increased numbers of small primary lysosomes and clear apical vesicles (fig. 1). Ultrastructural morphometry coupled with biochemical analysis of lysosomal

Table I. Chemical characteristics of mammalian MT

Molecular weight	6,800
Dominant amino acids	Cys, Gly, Ser, Lys, no aromatics
Sequence characteristics	Cys × Cys
Isomers	2 major isomers but many others
Isoelectric point(s)	4.1–4.5
Mole metal/mole protein	7
Clusters	2 of 4 and 3 metals each
Binding groups	4 SH: metal atom
Metals which bind and induce MT synthesis	Cd, Zn, Cu, Hg, Ag
Secondary structure	>95% random
Tissue distribution	widely distributed

cathepsin D activity have been used to quantitatively delineate the time-course of these events [16, 22] and data from these studies demonstrate an apparent disruption of normal lysosomal biogenesis prior to increased apical vesiculation. These data suggest that this phenomenon results from a disruption of normal pinocytotic vesicle-primary lysosome fusion within the renal proximal tubule cells. Functional studies [16, 22] demonstrated that these events were associated with the development of a tubular proteinuria and calciuria similar to that observed in humans and animals under chronic exposure conditions. All of the above effects were found to be prevented/attenuated by induction of the renal MT pool via Zn pretreatment which provided an available intracellular sink for Cd^{2+} ions released from reabsorbed/degraded CdMT complex. These studies suggest that MT is centrally involved in the mechanism(s) of Cd^{2+} nephrotoxicity by first acting as vehicle for the delivery of this metal to the target cell population and secondly by modulating Cd^{2+} intracellular toxicity of incoming Cd^{2+} ions via vacant or displacable binding sites. The exact mechanism by which Cd^{2+} ions not bound to MT disrupt the normal process of renal lysosome biogenesis noted above is presently not clear but may involve Cd^{2+} activation of calmodulin [26] prior to renal induction of MT. In any event, this process appears to be a primary cause for the tubular proteinuria. Other studies [21, 22] have shown that the calciuria observed in this model appeared to follow the tubular proteinuria and did not appear to be linked to a defect in Ca^{2+} membrane transport. The latter effect may also play a role in the development of osteomalacia after Cd^{2+} exposure.

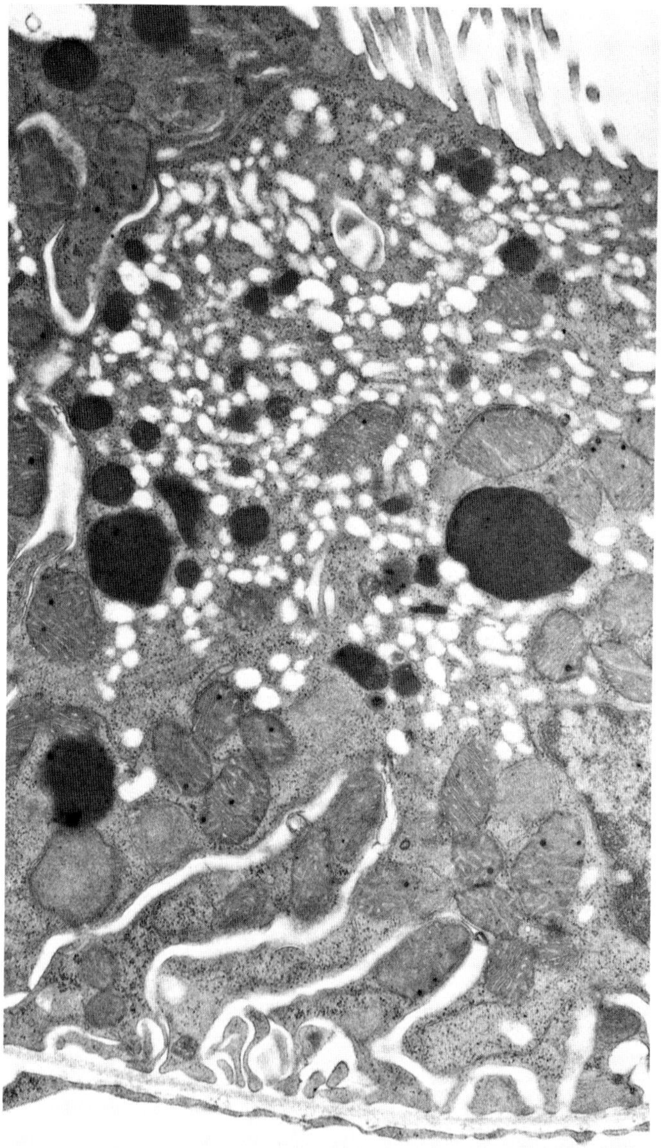

Fig. 1. Electron micrograph of a kidney proximal tubule cell following parenteral administration of CdMT showing apical vesiculation and small electron-dense lysosomes. ×14,750.

Table II. Chemical characteristics of renal PbBP

Molecular weight	8,300
Dominant amino acids	Asp, Glu, Gly Phe, Tyr
Sequence	in progress
Isomers	not apparent
Isoelectric point	5.3
Metal, atoms/mole	unknown
Clusters	unknown
Metal binding group	apparently COOH
Inducible synthesis	not apparent
Secondary structure	unknown
Tissue distribution	kidney and urine only

Renal Lead-Binding Proteins

In contrast with MT, the more recently discovered [27–32] rat renal PbBP are less well characterized. Table II gives the currently known chemical and immunological characteristics of the low molecular weight protein from rat kidney. This cytosolic PbBP is somewhat larger in size (8,300 daltons) than MT and differs in a number of other characteristics such as amino acid composition, isolectric point, chromatographic characteristics and lack of immunological cross-reactivity with MT by Western Blot analysis [32]. In addition, recent Western Blot analyses have shown that this protein is highly specific to the kidney and urine [Duval, unpublished observations] and that polyclonal antibodies to this molecule also do not cross-react with MT or the PbBP from rat brain. ^{203}Pb-binding studies [28] using partially purified renal PbBP have previously demonstrated that it has dissociation constant (K_d) for Pb of 10^{-8} *M* indicating a relatively high affinity. These studies also demonstrated that PbBP facilitates the cell-free nuclear translocation of Pb suggesting that it may play a central role in the formation of Pb-induced inclusion bodies (fig. 2) and attendant changes in renal gene expression [35–38]. Other studies [28–29] demonstrated that Cd^{2+} and Zn^{2+} but not Ca^{2+} were highly effective competitors against Pb with regard to binding. Cd^{2+} also markedly inhibited uptake of ^{203}Pb into kidney nuclei while Zn^{2+} actually increased uptake.

These data may also, in part, explain the previously reported [39] inhibition of Pb intranuclear inclusion body formation in animals receiv-

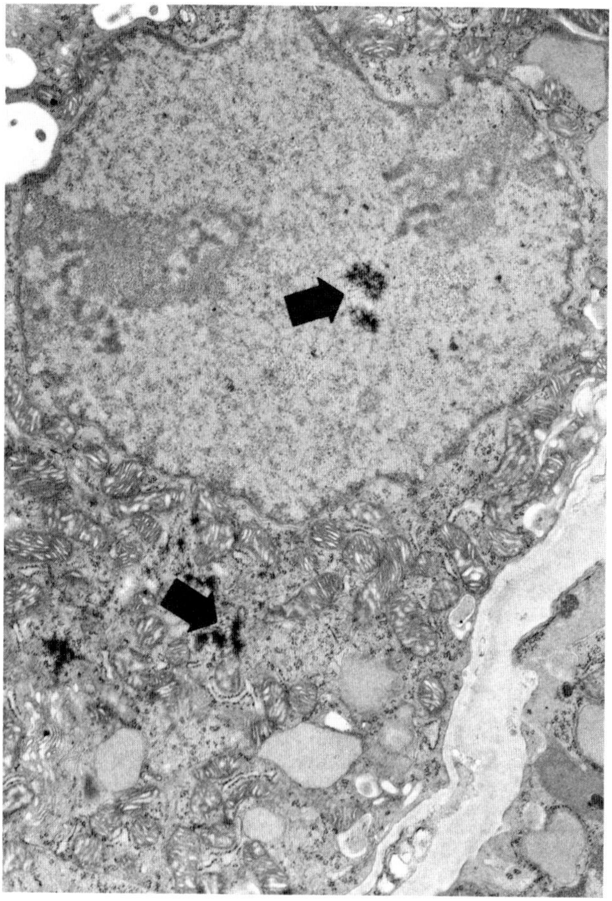

Fig. 2. Electron micrograph of a kidney proximal tubule cell following an intravenous dose of Pb showing both cytoplasmic and intranuclear inclusion bodies (arrows). ×9,300.

ing a combination of Pb and Cd in the diet for 10 weeks. The above data suggest that renal PbBP shares a number of characteristics (tissue specificity, high affinity, limited binding capacity, relative ligand specificity, ability to facilitate nuclear movement of the ligand) with 'receptors' for other types of biologically active substances [40]. Molecular biology studies are currently in progress to evaluate the final receptor criterion (correlation with biological response) for this protein to determine if our overall hypothesis is correct with respect to its possible role in mediating the previously reported [35–38] highly specific Pb-induced changes in renal gene expression.

In addition to the possible roles in mediating Pb-induced intranuclear effects, PbBP and MT have been shown [2, 30–32] to attenuate the effects of Pb on sensitive target molecules in the heme biosynthetic pathway, such as δ-aminolevulinic acid dehydratase (ALAD), which is remarkably resistant to Pb inhibition following chronic in vivo exposure [41, 42]. The chemical mechanism of this protection appears to involve both Pb chelation by these proteins and Zn donation with subsequent activation of this Zn-dependent enzyme.

More recent studies [43] have demonstrated that for MT, the capacity to protect the enzyme from Pb inhibition is highly dependent upon the number of Zn atoms/molecule and that a decreased protective capacity is observed when Cd atoms are present on the MT. It should be noted that while the PbBP may be of value in preventing Pb inhibition of ALAD, it may actually enhance Pb inhibition of other heme biosynthetic enzymes, such as those localized in the mitochondria by facilitating the transmembrane movement of this toxin.

Directions for Future Research

Results of the above studies suggest that common toxic metals such as Cd and Pb which are accumulated by the kidney may also interact with each other. For both of these metals, high affinity metal-binding proteins such as MT or PbBP appear to play major roles in mediating these interactions. A state of competition exists between binding sites on high affinity proteins for both essential and toxic metals in the kidney and presumably other tissues as well. While most metal-binding proteins such as MT or PbBP are relatively specific for certain metals, they are also capable of accepting other metals into vacant or displacable binding sites. Studies showing MT activation of Zn/Cu metalloenzymes [2–5] suggest that this high-affinity protein is capable of donating these metals to enzymes with ostensibly lower metal-binding affinities providing further evidence for the apparent dynamic nature of metal-protein interactions. Furthermore, recent studies from this laboratory also suggest that high affinity proteins such as MT and PbBP may act in concert to mediate the effects of a toxic metal such as Pb on a sensitive target molecule such as

ALAD by simulateneously regulating the bioavailability of a toxic metal (Pb) and an essential metal (Zn) to this enzyme.

Another interesting aspect of metal-induced nephrotoxicity concerns the mechanism by which these agents interact with the cellular genome to induce the synthesis of metal-binding proteins such as MT, renal Pb-inclusion body protein or other metal stress proteins [37, 38]. Studies from this laboratory suggest that the endogenous renal PbBP is capable of acting in a manner similar to other receptor molecules to facilitate this intranuclear interaction. This would provide a viable molecular mechanism for explaining why only certain cells in the kidney are specifically affected by Pb and other metals and perhaps how metals such as Pb produce adenocarcinomas in renal proximal tubule cell populations.

References

1 Kagi, J.H.R.; Nordberg, M.: in Metallothionein (Birkhäuser, Basel 1979).

2 Goering, P.L.; Fowler, B.A.: Mechanism of kidney metallothionein reversal of lead inhibition of δ-aminolevulinic acid dehydratase. Archs Biochem. Biophys. *253:* 48–56 (1987).

3 Udom, A.O.; Brady, F.O.: Reactivation in vitro of zinc-requiring apo-enzymes by rat liver zinc-thionein. Biochem. J. *187:* 329–338 (1980).

4 Li, T.Y.; Kraker, A.J.; Shaw, C.F.; Petering, D.H.: Ligand substitution reactions of metallothionein with EDTA and apocarbonic anhydrase. Proc. natn. Acad. Sci. USA *77:* 6334–6338 (1980).

5 Geller, B.L.; Winge, D.R.: Metal-binding sites of rat liver Cu-thionein. Archs Biochem. Biophys. *213:* 109–117 (1982).

6 Chang, C.C.; Vander Mallie, R.J.; Garvey, J.S.: A radioimmunoassay for human metallothionein. Toxicol. appl. Pharmacol. *55:* 94–102 (1980).

7 Garvey, J.S.; Chang, C.C.: Detection of circulating metallothionein in rats injected with zinc or cadmium. Science *214:* 805–807 (1981).

8 Cherian, M.G.; Shaikh, Z.A.: Metabolism of intravenously injected cadmium-binding protein. Biochem. biophys. Res. Commun. *65:* 863–869 (1975).

9 Cherian, M.G.; Goyer, R.A.; Delaquerriere-Richardson, L.: Cadmium-metallothionein-induced nephrotoxicity. Toxicol. appl. Pharmacol. *38:* 399–408 (1976).

10 Foulkes, E.C.: Renal tubular transport of cadmium-metallothionein. Toxicol. appl. Pharmacol. *45:* 505–512 (1978).

11 Squibb, K.S.; Ridlington, J.W.; Carmichael, N.G.; Fowler, B.A.: Early cellular effects of circulating cadmium-thionein on kidney proximal tubules. Environ. Health Perspect. *28:* 287–296 (1979).

12 Squibb, K.S.; Pritchard, J.B.; Fowler, B.A.: The renal metabolism and toxicity of metallothionein; in Foulkes, Biological roles of metallothionein. Proc. USA-Japan Workshop on Metallothionein, pp. 181–192 (Elsevier/North-Holland, Amsterdam 1982).

13 Squibb, K.S.; Fowler, B.A.: Intracellular metabolism and effects of circulating cadmium metallothionein in the kidney following acute treatment. Environ. Health Perspect. *54:* 31 (1984).
14 Cain, K.; Holt, D.E.: Studies of cadmium thionein induced nephropathy. Time course of cadmium thionein uptake and degradation. Chem.-Biol. Interact. *43:* 223–237 (1983).
15 Suzuki, K.T.: Studies of cadmium uptake and metabolism by the kidney. Environ. Health Perspect *54:* 21–30 (1984).
16 Squibb, K.S.; Pritchard, J.B.; Fowler, B.A.: Cadmium metallothionein nephropathy. Ultrastructural/biochemical alterations and intracellular cadmium binding. J. Pharmac. exp. Ther. *229:* 311–321 (1984).
17 Wall, D.A.; Maack, T.: Endocytic uptake on transport and catabolism of proteins by epithelial cells. Am. J. Physiol. *248:* C12–C20 (1985).
18 Adams, R.G.; Harrison, J.F.; Scott, P.: The development of cadmium induced proteinuria, impaired renal function and osteomalacia in alkaline battery workers. Q. Jl Med. *38:* 425–436 (1969).
19 Kanzantzis, G.: Renal tubular dysfunction and abnormalities of calcium metabolism in cadmium workers. Environ. Health Perspect. *28:* 155–159 (1979).
20 Kjellstrom, T.: Renal effects; in Friberg, Elinder, Kjelstrom, Nordberg, Cadmium and health: a toxicological and epidemiological appraisal, pp. 47–86 (CRC Press, Boca Raton 1986).
21 Goering, P.L.; Squibb, K.S.; Fowler, B.A.: Increased renal calcium excretion during cadmium-induced proximal tubule cell injury; in Hemphill, Trace substances in environmental health, XIX, pp. 22–36 (University of Missouri Press, Columbia 1986).
22 Fowler, B.A.; Goering, P.L.; Squibb, K.S.: Mechanism of cadmium-metallothionein-induced nephrotoxicity. Relationship to altered renal calcium metabolism; in Kagi, Metallothionein. 2nd Int. Meet. Metallothionein and other Low Molecular Weight Metal-Binding Proteins pp. 661–668 (Birkhäuser, Basel 1987).
23 Fowler, B.A.; Nordberg, G.F.: The renal toxicity of cadmium-metallothionein. Morphometric and x-ray microanalytical studies. Toxicol. appl. Pharmacol. *46:* 609–623 (1978).
24 Webb, M.; Etienne, A.T.: Studies on the toxicity and metabolism of cadmium thionein. Biochem. Pharmacol. *26:* 25–30 (1977).
25 Nordberg, G.F.; Goyer, R.; Nordberg, M.: Comparative toxicology of cadmium metallothionein and cadmium chloride on mouse kidney. Archs Path. *99:* 192–197 (1975).
26 Chao, S.H.; Suzuki, Y.; Zysk, J.R.; Cheung, W.Y.: Activation of calmodulin by various metal cations as a function of ionic radius. Molec. Pharmacol. *26:* 75–82 (1984).
27 Oskarsson, A.; Squibb, K.S.; Fowler, B.A.: Intracellular binding of lead in the kidney. Partial isolation and characterization of post-mitochondrial supernatant lead-binding components. Biochem. biophys. Res. Commun. *104:* 290–298 (1982).
28 Mistry, P.; Lucier, G.W.; Fowler, B.A.: High affinity lead-binding proteins from rat kidney cytosol. Mediate cell-free nuclear translocation of lead. J. Pharmac. exp. Ther. *232:* 462–469 (1985).

29 Mistry, P.; Mastri, C.; Fowler, B.A.: Influence of metal ions on renal cytosolic lead-binding proteins and nuclear uptake of lead in the kidney. Biochem. Pharmacol. *35:* 711–713 (1986).

30 Goering, P.S.; Fowler, B.A.: Regulation of lead inhibition of δ-aminolevulinic dehydratase by a high affinity renal lead-binding protein. J. Pharmac. exp. Ther. *231:* 66–71 (1984).

31 Goering, P.L.; Fowler, B.A.: Mechanisms of renal lead-binding protein protection against lead-inhibition of delta-aminolevulinic acid dehydratase. J. Pharmac. exp. Ther. *234:* 365–371 (1985).

32 Goering, P.L.; Mistry, P.; Fowler, B.A.: A high affinity lead-binding protein in brain attenuates lead inhibition of δ-aminolevulinic acid dehydratase. Comparison with a renal lead-binding protein. J. Pharmac. exp. Ther. *237:* 220–225 (1986).

33 DuVal, G.E.; Fowler, B.A.: Purification of the low molecular weight non-metallothionein lead-binding protein (PbBP) from rat kidney. Toxicologist *7:* 73 (1987).

34 Gilg, D.E.O.; Pentecost, B.; DuVal, G.E.; Fowler, B.A.: Antibody specificity to the low molecular weight lead-binding protein. Fed. Proc. (1987).

35 Fowler, B.A.; Mistry, P.; Victery, W.W.: Ultrastructural morphometric studies of lead inclusion body formation in kidney proximal tubule cells. Relationship to altered renal protein synthetic patterns. Toxicologist *5:* 53 (1985).

36 Mistry, P.; Mastri, C.; Fowler, B.A.: Lead-induced alterations of renal gene expression within subcellular compartments. Toxicologist *7:* 78 (1987).

37 Shelton, K.R.; Egle, P.M.: The proteins of lead-induced intranuclear inclusion bodies. J. biol. Chem. *257:* 11802–11807 (1982).

38 Shelton, K.R.; Todd, J.M.; Egle, P.M.: The induction of stress-related proteins by lead. J. biol. Chem. *261:* 1935–1940 (1986).

39 Mahaffey, K.R.; Capar, S.G.; Gladen, B.C.; Fowler, B.A.: Concurrent exposure to lead, cadmium and arsenic. Effects on toxicity and tissue metal concentrations in the rat. J. Lab. clin. Med. *98:* 463–481 (1981).

40 Clark, J.H.; Peck, E.J., Jr.: Steroid hormone receptors: basic principles and measurement; in Schrader, O'Malley, Laboratory methods manual for hormone action and molecular endocrinology; 6th ed., pp. 1-1, 1-56 (Biological Associates, Houston 1982).

41 Fowler, B.A.; Kimmel, C.A.; Woods, J.S.; McConnell, E.E.; Grant, L.D.: Chronic low-level lead toxicity in the rat. III. An integrated assessment of long-term toxicity with special reference to the kidney. Toxicol. appl. Pharmacol. *56:* 59–77 (1980).

42 Oskarsson, A.; Fowler, B.A.: Effects of lead on the heme biosynthetic pathway in rat kidney. Exp. molec. Path. *43:* 397–408 (1985).

43 Goering, P.L.; Fowler, B.A.: Metal constitution of metallothionein influences inhibition of δ-aminolevulinic acid dehydratase by lead. Biochem. J. *245:* 339–345 (1987).

Bruce A. Fowler, PhD, Toxicology Program, University of Maryland,
660 West Redwood Street, Baltimore, MD 21201 (USA)

Contr. Nephrol., vol 64, pp. 93–101 (Karger, Basel 1988)

Ultrastructural Studies in Experimental Lead Intoxication[1]

E. Bonucci, G. Silvestrini

Department of Human Biopathology, Section of Pathological Anatomy, 'La Sapienza' University, Rome, Italy

Lead intoxication, or saturnism, has long been known, at least since the Roman Empire, and mention can be already found in the ancient literature of the so-called 'colica pittonum', a severe gastroenteritis with neurological disorders, chiefly due to ingestion of wine treated with antifermentative lead salts [Eisinger, 1982]. It has even been suggested that the decline of the Roman Empire was partly due to the chronic lead intoxication consequent to the use of drinking water from lead pipes.

In spite of this long history, only the most conspicuous pathologic changes of lead intoxication, such as those which occur in neurons, blood cells, and epithelial cells of renal tubules [Hass et al., 1964] have been studied in detail. Other tissues have been less completely studied although lead can damage most of them, because it is absorbed in the intestine, transported to the liver and, once entered the systemic circulation, incorporated in tissues, especially in bone matrix.

One of the best known changes induced by lead in tissues is represented by the development of inclusion bodies in the nuclei of neurons, hepatocytes and, above all, of epithelial cells of the renal tubuli [Angevine et al., 1962; Beaver, 1961; Richter et al., 1968]. Because these bodies are completely similar to those in osteoclasts, they will be considered in connection with these bone cells.

The other ultrastructural changes which can be found in soft tissues have been rarely studied. On the other hand, it must be conceded that these changes are neither conspicuous nor constant and consequently are

[1] This investigation has been supported by grants of the Consiglio Nazionale delle Ricerche and the Ministero della Pubblica Istruzione.

difficult to find. In order to obtain further data on them, and to verify the results previously obtained from histological, histochemical and ultrastructural investigations of tibiae and femurs of lead-intoxicated collie puppies [Bonucci et al., 1983], 20 rats were supplied for 20 days with drinking water containing lead (50 mg lead acetate in 250 ml of water, renewed every day). Morphological studies were undertaken on small intestine, liver, lung, myocardium, brain and bone specimens. Preliminary results of this investigation have been reported previously [Silvestrini and Bonucci, 1985].

In the small intestine of lead-intoxicated rats the morphology of the epithelial cells appeared completely normal, although they absorb the element during digestion. The microvilli were perfectly preserved and the cell organelles normally structured. However, at high enlargement, very small, electron-dense granules, not found in controls, were seen on the inner surface of the membranes of the microvilli. These granules were visible also in unstained sections, although they were less electron-dense than in stained sections. They were not numerous and were not present in all the epithelial cells. Similar electron-dense granules were contained in cytoplasmic vacuoles, probably lysosomes, of the epithelial intestinal cells and in expansions of the intercellular clefts, below the terminal bar (fig. 1). These vacuoles and expansions were also present in controls, but they did not contain any electron-dense granules.

At present, it is not possible to say if these granules contain lead, because on account of their smallness their electron-probe microanalysis has been unsuccessful. However, the possibility that the granules represent aggregates of lead on the way of being absorbed by the intestinal cells is supported by the fact that they were not present in controls.

It is possible that the cells were not altered by the presence of the element, because it is confined to vacuoles and not directly involved in the cell metabolism.

In the liver, however, the element probably interferes with the cell activity. This is chiefly suggested by the presence of inclusion bodies in the nuclei of a few hepatocytes. The cells without inclusion bodies appeared completely normal. However, some of them contained amorphous cytoplasmic vacuoles, which probably corresponded to lipid droplets, whose membranes were covered by very small granules. Similar granules were found in mitochondria.

The study of the nervous tissue has not yet been completed. The most evident change was the presence of inclusion bodies in the nucleus of a

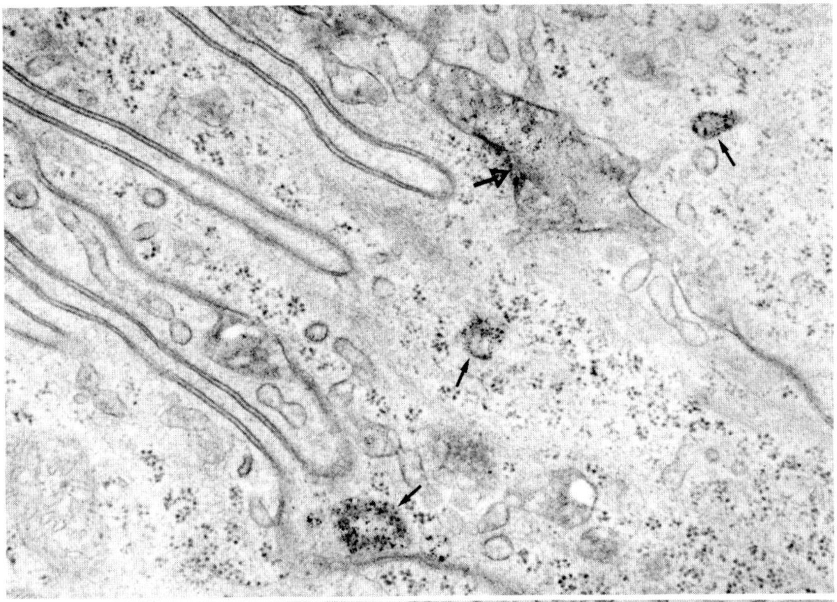

1

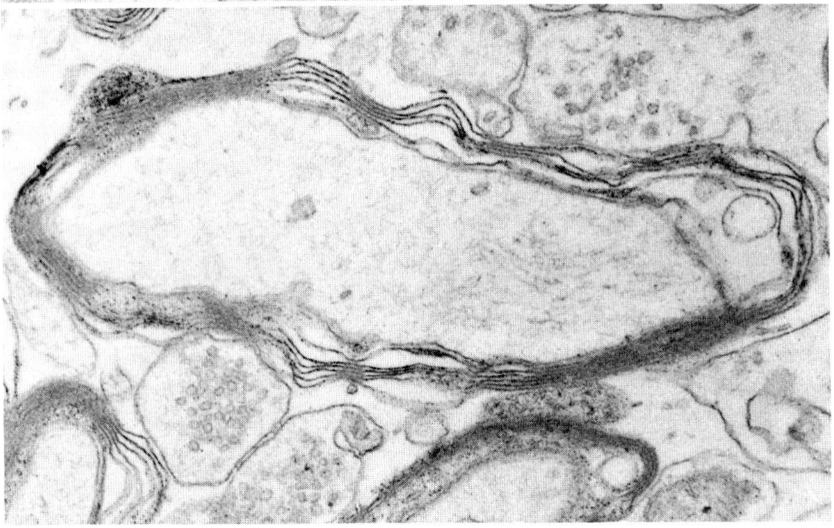

2

Fig. 1. Detail of the cytoplasm of epithelial intestinal cells: aggregates of electron-dense granules are visible in vacuoles (black arrows) and in expansion of the intercellular clefts (open arrow). Uranyl acetate and lead citrate. ×38,500.

Fig. 2. Degenerated and partially dissociated myelin sheath: note electron-dense granules in the swollen segments of the sheath. Uranyl acetate and lead citrate. ×50,500.

few neurons. In addition, there were myelin degenerative changes consisting of focal areas of swelling and dissociation. These changes might be due to artifacts of preparation. However, most of the swollen and dissociated areas contained the same electron-dense granules already found in intestinal and hepatic cells, and this suggests that they represent lead granules accumulated in myelin degenerative areas (fig. 2).

Similar granules were occasionally present in lipid vacuoles of the type I alveolar cells of the lung and in mitochondria of the myocardium.

It is well known that lead is incorporated in bone matrix during calcification [Scheiman-Tagger and Brodie, 1964]: the element is transported through the osteoblasts and is incorporated into the hydroxyapatite crystals at the calcification front, so that it becomes a permanent component of the bone matrix until this is reabsorbed by the osteoclasts.

In spite of this incorporation, the structure of the metaphyseal bone did not appear greatly changed in lead-intoxicated rats. The only apparent abnormality was an increase of the number of the bone trabeculae, so that the metaphysis showed an initial condition of osteosclerosis. The growth zone of the cartilage was unchanged and the maturation of the chondrocytes was normal. The penetration of capillary vessels into the lacunae of hypertrophic and degenerate chondrocytes was also normal.

The only change appreciable under the light microscope in paraffin sections of metaphyseal bone was represented by the fact that the osteoclasts were more numerous than in controls and that most of them contained a lot of nuclei. However, on semithin sections most of these osteoclasts were abnormal. In fact, they contained dense, deeply stained intranuclear and intracytoplasmic granules collected in small clusters.

The osteoclasts with these inclusion bodies had many nuclei, part of which were shrunken and condensed. In spite of the presence of these nuclei, and of the inclusion bodies in the nuclei themselves and in the cytoplasm, most of the osteoclasts had a normal configuration, were attached to the bone matrix, and appeared to be normally engaged in bone resorption. Others, however, were detached from the bone matrix, were more or less dislocated in the central zones of the medullary spaces, and appeared no longer engaged in bone resorption.

Under the electron microscope, the osteoclasts with inclusion bodies often appeared completely normal. In particular, inorganic crystals disaggregated from the bone matrix could be recognized between the infoldings of developed ruffled borders and in cytoplasmic vacuoles, showing that the resorption activity of these osteoclasts continued, in spite of the

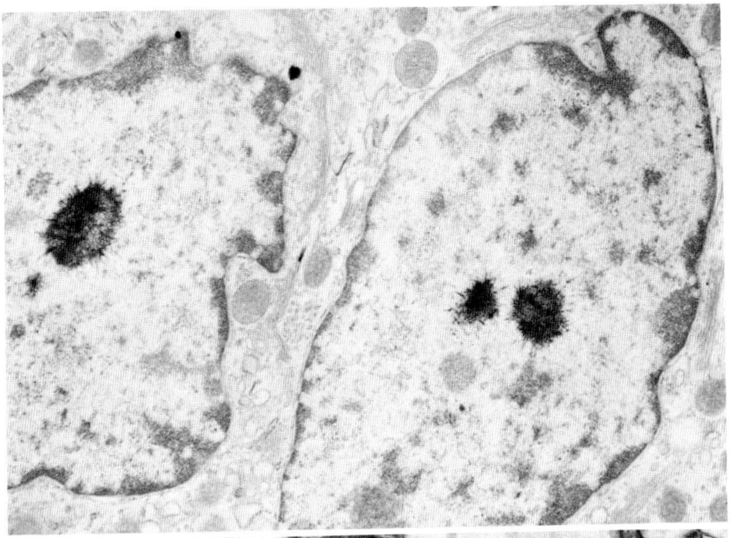
3

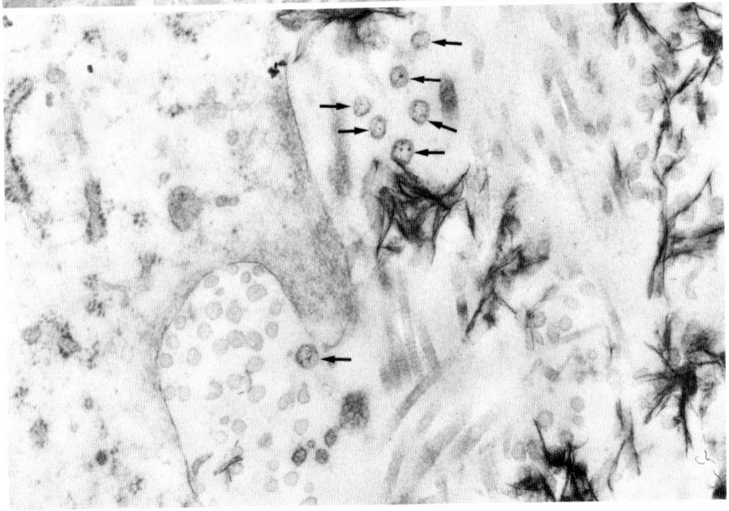
4

Fig. 3. Osteoclast nuclei containing inclusion bodies. Uranyl acetate and lead citrate. ×12,600.

Fig. 4. Small vesicles (arrows) containing electron-dense granules in the osteoid border of a lead-intoxicated rat. Uranyl acetate and lead citrate. ×35,000.

presence of the inclusion bodies. These corresponded to electron-dense aggregates of fibrillary and granular structures, resembling clusters of phagocytosed crystals, and were present not only in the cytoplasm but also in the nuclei (fig. 3). Similar inclusion bodies have been described in osteoclasts by Hsu et al. [1973] and by Van Mullem and Stadhouders [1974].

In many cases, two or more inclusion bodies were contained in the same nucleus. They appeared as roundish aggregates of needle- and filament-like structures which often had a radiating arrangement (fig. 3). On the whole, these aggregates were similar to the inclusion bodies already described in the cells of the kidney, liver and nervous tissue.

The same type of inclusion bodies was found in the cytoplasm. Again, they resembled phagocytosed crystals; however, they were not contained in lysosomes. Moreover, they resisted decalcification with EDTA or formic acid, although their electron density was greatly reduced by this treatment.

At high enlargement, both the inclusion bodies found in the nuclei and those found in the cytoplasm consisted of filaments and granules. Granules were also present around them and within mitochondria, as reported above for hepatocytes, lung cells and myocardium.

Electron-probe microanalysis of these aggregates of granules and filaments performed on unstained preparations has shown that they contain lead and that the amount of this element is roughly proportional to the concentration of the granules [Bonucci et al., 1983].

It is interesting to note that the inclusion bodies of the osteoclasts can be stained by silver nitrate. With this staining method, which increases the histological evidence of the inclusion bodies, it is easy to see that they were contained also in osteoclasts which were no longer in direct contact with the bone matrix. Although these cells might be considered bone marrow macrophages, their shape and the presence of inclusion bodies in them suggest that they represent lead-damaged osteoclasts, detached from the bone matrix they were reabsorbing and probably degenerating and dying. The fate of the normal osteoclasts is still unknown [Bonucci, 1981]; shrinkage of their nuclei, detachment from the bone matrix, and fragmentation of the cytoplasm with dispersion of fragments in the bone marrow might represent the ultimate fate of the lead-intoxicated osteoclasts as well as that of the normal osteoclasts.

The inclusion bodies of the osteoclasts can be stained also by silver nitrate after periodic acid oxidation, a technique which shows glycoproteins. The presence of these substances in the inclusion bodies was confirmed under the electron microscope by their staining with acidic phosphotungestic acid, another technique which stains molecules containing vic-glycol groups. It is interesting that with this staining method the central part of the inclusion bodies sometimes appeared almost empty.

The mechanism of formation of the inclusion bodies is still unknown. Choie et al. [1975] have shown that the protein synthesis is required for their formation, because in renal cells their development is inhibited by cycloheximide administration. In these cells, Richter et al. [1968] have shown that the inclusion bodies contain nonhistonic proteins. Moreover, the same authors have found by biochemical analysis of partly purified inclusion bodies isolated by cell fractionation that they consist of a lead-protein complex. The histochemical and ultrastructural results confirm the presence of organic material in these bodies and show that it consists at least in part of glycoproteins.

Of the other bone cells, osteocytes and osteoblasts appeared normally structured, even when they were placed in proximity of the osteoclasts with inclusion bodies. Also the calcification front appeared normally structured, with many calcification centers between and on the collagen fibrils of the osteoid border. However, many small vesicles (fig. 4) were often present between the uncalcified collagen fibrils in proximity of early aggregates of crystals and calcification centers. Most of these vesicles contained very small, electron-dense granules, which were only partly solubilized by decalcifying solutions. Although these granules have not been examined by electron probe microanalysis, they probably contain lead, because they were completely similar to the lead-containing granules found in osteoclasts [Bonucci et al., 1983] and in odontoblast vesicles [Ozawa et al., 1981] of lead-intoxicated rats. On the other hand, these vesicles were similar to the so-called 'matrix vesicles' [Anderson, 1976; Bonucci, 1984] which were present in osteoid borders of the controls.

The interest of matrix vesicles is principally due to the fact that they represent the locus of initial calcification in cartilage and bone, as shown by the presence inside them of early inorganic crystals [Anderson, 1976; Bonucci, 1967, 1984]. In this connection, it is possible that the vesicles found in lead intoxicated rats were the same structures as the matrix vesicles found in controls. However, they were more numerous than in controls and contained electron-dense granules. It is attractive to speculate that they were normal matrix vesicles produced at an increased rate for exporting lead from the osteoblast cytoplasm into the extracellular matrix.

In conclusion, the available results show that the principal damage induced by lead intoxication is represented by degenerative changes, especially of myelin sheaths, and by the formation of inclusion bodies in renal epithelial cells, hepatocytes, neurons, and osteoclasts. Probably, the inclusion bodies have the same structure and composition in all these cells. They have a glycoproteic component and contain lead. They form only in the cells which accumulate the element in their cytoplasm, while they are not present in cells, as intestinal cells and osteoblasts, in which the lead transit is rapid. Lead accumulation in osteoclasts certainly occurs during bone matrix resorption, when the lead-contaminated apatite crystals are removed from the bone matrix, are collected into lysosomes, and are eventually solubilized, with release of the lead they contain. The osteoclasts with the inclusion bodies undergo nuclear shrinkage and pyknosis, detach from the bone matrix, and probably die. It is possible that the soft tissue cells which contain inclusion bodies have the same dramatic fate.

The presence of vesicles containing electron-dense granules in proximity of the osteoblasts of lead-intoxicated rats seems of particular interest. Assuming that these granules contain lead, it could be speculated that the vesicles represent a way of exporting the element to the outside of the osteoblasts. The formation of the vesicles might be induced by the necessity of decreasing the intracellular concentration of lead, and confining the element to close compartments. If these hypotheses are correct, it is possible to speculate that also the normal production of matrix vesicles is dependent on a vesiculation process induced by excessive intracellular concentration of calcium ions and by the consequent need of exporting it to the extracellular space. Consequently, lead could be a trace element useful for the study of the normal mechanism of matrix vesicle production and of the calcification process.

References

Anderson, H.C.: Matrix vesicles of cartilage and bone; in Bourne, The biochemistry and physiology of bone, vol. 4, pp. 135–157 (Academic Press, New York 1976).

Angevine, J.M.; Kappas, A.; DeGowin, R.; Spargo, B.H.: Renal tubular inclusions of lead poisoning. A clinical and experimental study. Archs Path. *73:* 486–494 (1962).

Beaver, D.L.: The ultrastructure of the kidney in lead intoxication with particular reference to intranuclear inclusions. Am. J. Path. *39:* 195–208 (1961).

Bonucci, E.: Fine structure of early cartilage calcification. J. Ultrastruct. Res. *20:* 33–50 (1967).

Bonucci, E.: New knowledge on the origin, function and fate of osteoclasts. Clin. Orthop. *158:* 252–269 (1981).

Bonucci, E.: Matrix vesicles: their role in calcification; in Linde, Dentin and dentinogenesis, vol. 1, pp. 135–154 (CRC Press, Boca Raton 1984).

Bonucci, E.; Barckhaus, R.H.; Silvestrini, G.; Ballanti, P.; Di Lorenzo, G.: Osteoclast changes induced by lead poisoning (saturnism). Appl. Path. *1:* 241–250 (1983).

Choie, D.D.; Richter, G.W.; Young, L.B.: Biogenesis of intranuclear lead-protein inclusions in mouse kidney. Beitr. Path. *155:* 197–203 (1975).

Eisinger, J.: Lead and wine. Eberhard Gockel and the colica pictonum. Med. Hist. *26:* 486–494 (1982).

Hass, G.M.; Brown, D.V.L.; Eisenstein, R.; Hemmens, A.: Relations between lead poisoning in rabbit and man. Am. J. Path. *45:* 691–727 (1964).

Hsu, F.S.; Krook, L.; Shively, J.N.; Duncan, J.R.; Pond, W.G.: Lead inclusion bodies in osteoclasts. Science *181:* 447–448 (1973).

Ozawa, H.; Yamada, M.; Yamamoto, T.: Ultrastructural observations on the location of lead and calcium in the mineralizing dentine of rat incisors; in Ascenzi, Bonucci, De Bernard, Matrix vesicles, pp. 179–187 (Wichtig, Milan 1981).

Richter, G.W.; Kress, Y.; Cornwall, C.C.: Another look at lead inclusion bodies. Am. J. Path. *53:* 189–217 (1968).

Scheiman-Tagger, E.; Brodie, A.G.: Lead acetate as a marker of growing calcified tissues. A modified method. Anat. Rec. *150:* 435–439 (1964).

Silvestrini, G.; Bonucci, E.: Ultrastructural study of bone and small intestine cells in lead-intoxicated rats; in Priest, Metals in bone, pp. 419–426 (MTP Press, Lancaster 1985).

Van Mullem, P.J.; Stadhouders, A.M.: Bone marking and lead intoxication. Early pathological changes in osteoclasts. Virchows Arch. Abt. B Zellpath. *15:* 345–350 (1974).

E. Bonucci, MD, Dipartimento di Biopatologia Umana, Sezione di Anatomia Patologica, Policlinico Umberto I, Viale Regina Elena 324, I-00161 Roma (Italia)

Contr. Nephrol., vol. 64, pp. 102–108 (Karger, Basel 1988)

Bone Lead and the Diagnosis of Lead Nephropathy

R.P. Wedeen [a,b,c], *F.L. Van de Vyver* [b], *P.C. D'Haese* [b], *W.J. Visser* [d], *M.M. Elseviers* [b], *L.J. Knippenberg* [b], *L.V. Lamberts* [b], *M.E. De Broe* [b], *V. Batuman* [a,c], *G. Schidlovsky* [d], *K. Jones* [e]

[a]Veterans Administration Medical Center, East Orange, N.J., USA; [b]University Hospital of Antwerp, Wilrijk and the University of Antwerp (UIA), Antwerp, Belgium; [c]University of Medicine and Dentistry of New Jersey, The New Jersey Medical School, Newark, N.J., USA; [d]University Hospital of Utrecht, Utrecht, The Netherlands; [e]Brookhaven National Laboratories, Upton, N.Y., USA

Recognition of lead nephropathy as a distinct clinical entity in the absence of the classical symptoms of acute lead poisoning was made possible by the use of the ethylenediaminetetraacetic acid (EDTA) lead mobilization test [1]. Chelatable lead has been widely accepted as a measure of body lead stores [2]. Since over 90% of the body lead burden is found in bone where it has a biological half-life of several decades, the blood lead concentration is an inadequate measure of cumulative lead absorption. Use of the EDTA lead mobilization test has shown that some patients previously thought to have renal disease due to gout or hypertension are, in fact, victims of excessive lead absorption [3–8]. Kidney disease has emerged as the most important sequel of chronic, low dose lead absorption in adults [9].

The validity of the EDTA test as a measure of cumulative lead stores in renal failure depends on a number of conditions. First, excessive mobilizable lead must not be a consequence of renal failure. This has been demonstrated in patients with renal disease of known, nonlead etiology, who were used as controls [1, 3–6]. Since excessive mobilizable lead is not a consequence of renal failure, it can be considered a cause unless the control groups are misleading. The control patients must, by definition, have proven renal diagnoses other than lead nephropathy.

The validity of the EDTA test as a measure of bone lead stores has gained support from recent measurements made in bone biopsies in Ant-

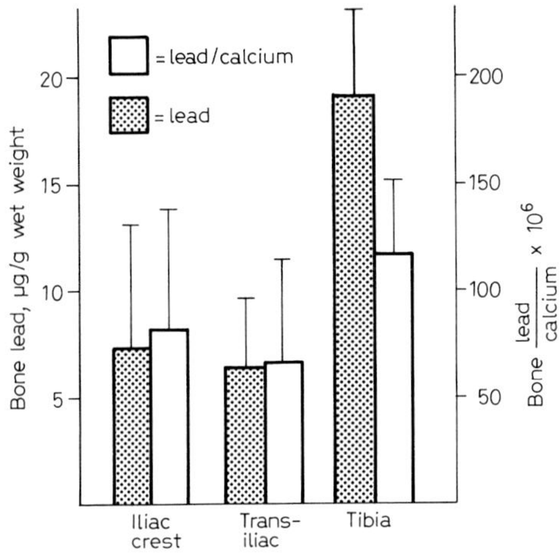

Fig. 1. Comparison of bone lead (▩) and Pb:Ca ratios (□) in iliac crest, transiliac and tibial biopsies from cadavers.

werp. Lead and calcium were measured by atomic absorption spectrometry (AAS) in 153 transiliac bone biopsies previously obtained for a study of aluminum osteodystrophy in dialysis patients [10]. The transiliac biopsies had been obtained from French, German and Belgian patients. The frequency distributions for lead and Pb:Ca ratio in these bones were nonparametric, while the calcium distribution from the same specimens was gaussian. These data demonstrate that renal failure per se does not cause accumulation of lead in bone but that approximately 5% in the studied sample of the European dialysis population has elevations of bone lead stores. Of importance in validating the EDTA test are histomorphometric observations made on biopsies from 5 dialysis patients with high bone lead and 18 with low bone lead. Bone histomorphometry did not differ in the high and low lead groups. Lead absorption therefore does not appear to alter uremic osteodystrophy which could account for systematic differences in mobilizable lead. The absence of systematic bone lead accumulation in interstitial nephritis as compared to other types of end-stage renal disease was supported by observations in a subset of 10 dialysis patients with analgesic nephropathy established by history and objective evidence of papillary necrosis. In interstitial nephritis due to analgesic abuse, bone lead and Pb:Ca ratios are lower than the mean for the dialysis population as a whole. Interstitial nephritis by itself, therefore, does not cause accumulation of bone lead compared to other forms of renal disease.

Table I. Reliability of paired cadaver bone biopsy measurements

	Number of subjects	Number of duplicate values	Linear correlation coefficient r
L-R iliac crest [lead]	7		0.98
L-R iliac crest Pb:Ca ratio	7		0.97
L-R transiliac [lead]	12[a]		0.82
L-R transiliac Pb:Ca ratio	12[a]		0.97
L-R tibia [lead]	7		0.52
L-R tibia Pb:Ca ratio	7		0.75
Duplicate[b] iliac crest [lead]		14	0.91
Duplicate[b] iliac crest Pb:Ca ratio		12	0.96
Duplicate[b] transiliac [lead]		24[a]	0.94
Duplicate[b] transiliac Pb:Ca ratio		23[a]	0.98

[a]Including one dialysis patient with a history of lead exposure.
[b]Adjacent biopsies.

Table II. Mean transiliac bone [lead] and bone Pb:Ca ratio (±SD)

Group	n	[lead] µg/g	Pb:Ca $\times 10^{-6}$ g/g
Dialysis	153	5.5±4.6	53±49
Dialysis (top 5%)	8	20.6±2.7	190±42
Dialysis due to analgesic nephropathy	10	3.7±2.1	39±19
Cadavers	11	6.3±1.1	65±19
Lead workers	13	29.8±17.8	285±179

The reproducibility of the bone lead measurements was demonstrated in multiple biopsies taken from 12 cadavers. Linear correlation coefficients for the Pb:Ca ratio tended to be higher than for bone lead content (table I), indicating that the Pb:Ca ratio is more reliable than the absolute lead concentration in bone biopsies (fig. 1).

The validity of the EDTA test as a measure of bone lead was demonstrated in 24 subjects in whom both transiliac biopsies and EDTA tests were performed. Thirteen of the study group were lead workers with a mean ±SD chelatable lead of 2,344± 1,617 µg approximately four times

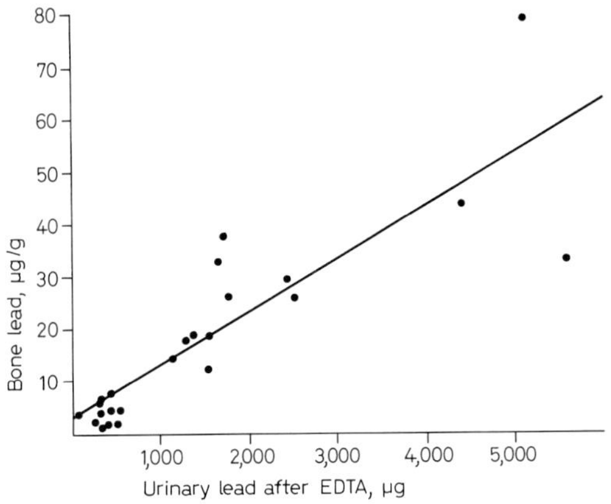

Fig. 2. Relation between the urinary lead excretion after EDTA and transiliac bone lead in 24 patients, including 13 lead workers and 11 unexposed individuals.

the level found in unexposed individuals. Urinary lead excretion was measured over one or 3 days (in normal and renal failure patients respectively) following the intramuscular injection of 2 g $CaNA_2EDTA$. The linear regression correlation coefficient for chelatable lead versus bone lead was 0.86 (fig. 2) and 0.83 versus the Pb:Ca ratio. The transiliac bone leads averaged 29.8 ± 18 μg/g in the lead workers and their Pb:Ca ratios averaged $285 \pm 179 \times 10^{-6}$ g/g (table II). The data from these lead workers place the bone biopsy measurements made in the dialysis patients in perspective. The mean bone lead and Pb:Ca ratios in the upper 5% of the European dialysis patients approach the levels found in lead workers. In contrast, the mean transiliac bone values for the 153 dialysis patients and the 11 unselected cadavers is about one fifth that found in the lead-exposed workers.

The Antwerp studies support the use of the EDTA lead mobilization test for the assessment of bone lead stores. In addition, they provide reference standards from which the diagnosis of excessive body lead burdens can be made from bone biopsies. However, the EDTA test is unsuitable for large population studies or for use in patients with end-stage renal disease. In order to determine the incidence of unsuspected excessive lead absorption in the general population, a noninvasive method that gives information comparable to that obtained with the EDTA test is needed. In vivo X-ray fluorescence (XRF) meets this need.

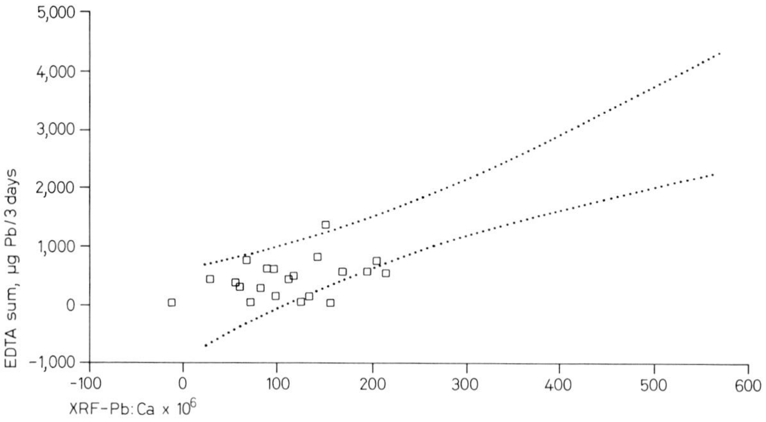

Fig. 3. X-ray fluorescence-Pb:Ca versus EDTA sum.

The Antwerp studies assist in the interpretation of in vivo measurement of bone lead by XRF (fig. 3). In eight cadavers, iliac crest, transiliac and tibial biopsies were obtained. Transiliac and iliac crest specimens give virtually identical results while the tibias showed distinctly higher lead concentrations, in part reflecting higher bone density and less trabecular bone (fig. 1). The mean tibial ratio is 1.75 times that of the ilium. We have therefore used this factor to compare transiliac measurements made by atomic absorption spectrometry to tibial bone lead measurements made by XRF.

XRF was induced with the 88 keV γ-ray emitted from a 30-mCi ^{109}Cd source placed in direct apposition to the anterior medial aspect of the central mid-diaphysial region of the tibia for 2,000s [11, 12]. The characteristic K X-rays in elastically scattered photons were measured with a germanium detector of 25 mm diameter. Measurement of the ratio of K X-rays to elastically scattered photons allows calculation of the Pb:Ca ratio from known photon cross-sections. This measurement is independent of geometry and soft tissues overlying the tibia. The uncertainty of the measurement (95% confidence limits) was 8% using phantoms containing 18 µg Pb/g and a Pb:Ca ratio of 150×10^{-6} g/g.

The XRF technique was validated by comparison with AAS measurements made in 6 amputated legs. The linear regression of XRF against the chemical measurements made in New Jersey gave a correlation coefficient of 0.91. AAS on split tibial biopsy samples from the same legs performed in Antwerp yielded nearly identical results.

Evidence that in vivo XRF can replace the EDTA lead mobilization test was obtained by comparing XRF Pb:Ca ratios to chelatable lead in 21 male patients. The linear regression obtained when comparing lead chelate excretion, to tibial XRF Pb:Ca ratios gave a correlation coefficient of 0.35 ($p < 0.001$). This relatively weak correlation can be explained for the fact that bone lead levels extended over only a very limited range. No lead workers were included in the study group. However, when the XRF data obtained in New Jersey is superimposed on the 95% confidence limits obtained from the EDTA and bone biopsy studies performed in Antwerp (fig. 3), the XRF measurements fit within the expected range.

Taken together, these studies demonstrate that XRF can replace the EDTA lead mobilization test for determining body lead stores in patients with kidney disease. Chelatable lead of 1,000 μg/3 days, typical for lead workers, gives an XRF Pb:Ca ratio of 180×10^{-6} g/g which corresponds to a tibial lead concentration of about 22 μg/g wet weight. This level of tibial lead can be measured with an error of 16% (95% confidence limits). The noninvasive XRF technique is suitable for estimating bone lead content at levels well below those found in lead workers.

These studies support the concept that the EDTA test assesses bone lead stores and indicate that bone lead content is not modified by renal failure per se. Only a small fraction of those with excessive body lead burdens develop renal disease and this fraction has not yet been determined. Tibial XRF is a practical, noninvasive tool for determining the frequency of elevated body lead stores and can be used to establish the risk of developing renal disease at different levels of body lead burdens.

References

1 Emmerson, B.T.: Chronic lead nephropathy. Kidney int. *4:* 1–5 (1973).
2 Inglis, J.A.; Henderson, D.A.; Emmerson, B.T.: The pathology and pathogenesis of chronic lead nephropathy occurring in Queensland. J. Path. *124:* 65–76 (1978).
3 Batuman, V.; Maeska, J.K.; Haddad, B.; et al.: The role of lead in gout nephropathy. New Engl. J. Med. *304:* 520 (1981).
4 Batuman, V.; Landy, E.; Maesaka, J.K.; et al.: Contribution of lead to hypertension with renal impairment. New Engl. J. Med. *309:* 17–21 (1983).
5 Craswell, P.W.; Price, J.; Boyle, P.D.; et al.: Chronic renal failure with gout: a marker of chronic lead poisoning. Kidney int. *26:* 319–323 (1984).

6 Craswell, P.W.; Price, J.; Boyle, P.D.; et al.: Chronic lead nephropathy in Queensland: alternative methods of diagnosis. Aust. N.Z. J. Med. *16:* 11–19 (1986).
7 Reynolds, P.P.; Knapp, M.J.; Baraf, H.S.B.; et al.: Moonshine and lead. Relationship to the pathogenesis of hyperuricemia in gout. Arthritis Rheum. *26:* 1057–1064 (1983).
8 Wright, L.F.; Saylor, R.P.; Cecere, F.A.: Occult lead intoxication in patients with gout and kidney disease. Rheumatology *11:* 517–520 (1984).
9 Wedeen, R.P.: Poison in the pot: the legacy of lead (Southern Illinois University Press, Carbondale 1984).
10 Wedeen, R.P.; D'Haese, P.; Verpooten, G.A.; et al.: Bone lead in dialysis patients. Abstracts. National Kidney Foundation, Inc., 15th Annual Scientific Meeting. Am. J. Kidney Dis. *1985:* A22.
11 Somervaille, L.J.; Laird, E.E.; Chettle, D.R.; et al.: Lead body stores assessed in vivo by x-ray fluorescence. Analyt. Chem. *55:* 40R–56R (1983).
12 Jones, K.W.; Kraner, H.W.; Schidlovsky, G.; et al.: in vivo determination of lead in the tibia by x-ray fluorescence. Biophys. J. *47:* 406a (1985).

M.E. De Broe, MD, PhD, University Hospital of Antwerp, Wilrijkstraat 10, B-2520 Edegem-Antwerp (Belgium)

Contr. Nephrol., vol. 64, pp. 109–112 (Karger, Basel 1988)

Pathophysiology of Aluminum-Induced Bone Disease

T. Drüeke [a], *M. Lieberherr* [b], *G. Cournot* [b]

[a]INSERM U 90 and [b]CNRS U.A. 583, Hôpital Necker-Enfants Malades, Paris, France

Aluminum-intoxicated patients may present with different patterns of osteopathy [1–4]. The type and severity of their bone disease depend on a variety of factors including the bone status before intoxication, the mode of aluminum overload and probably other, still unknown variables [2, 5, 6].

The complex interrelationship between aluminum and parathyroid hormone (PTH) has previously been addressed by us [1, 5] and others [2–4]. Suffice it to say that on the one hand, hyperparathyroidism appears to favor the tissue uptake of aluminum and to protect bone against the toxicity of aluminum, and on the other hand aluminum will suppress parathyroid function in the long run.

The interaction of aluminum with calcitriol (1,25-dihydroxy vitamin D_3) is still poorly understood. Thus, inhibitory [7] as well as stimulatory [8] effects of calcitriol on tissue accumulation of aluminum have been claimed in experimental animal models. Such discrepancies could be due to differences in dosage schedules of calcitriol and of aluminum and differences in other experimental designs. In uremic patients on dialysis, the short-term administration of 1α-$(OH)D_3$ leads to an increase in plasma aluminum concentration which could be the consequence of either increased intestinal absorption, increased bone resorption, or both [9]. The intestinal calcium absorption response to calcitriol is partially blocked by severe aluminum intoxication [10, 11], whereas bone calcium mobilization in response to calcitriol is not [10].

The effects of aluminum on bone in the absence of severe hyperparathyroidism include disturbed mineralization, severe low-turnover osteomalacia, and 'adynamic' or 'aplastic' bone disease. Aluminum not only affects bone mineralization but also bone formation [2–4, 12, 13]. In addition to reduced numbers of osteoblasts, low numbers of osteoclasts are also generally observed indicating decreased bone resorption. Even though most often bone and matrix formation fall together in response to aluminum, observations in the rat and in man indicate that a primary failure of osteoid synthesis may occur, indicating that collagen production may decrease without an associated defect in the mineralization of bone [2]. An uncoupling may also be observed between bone formation and resorption whereby the effects of aluminum on bone formation exceed those on bone resorption [13]. The mechanism by which aluminum induces these changes in bone dynamics is not fully understood.

We have previously shown that in organ culture, aluminum interferes directly not only with the synthesis of bone alkaline phosphatase, which reflects osteoblast activity, but also with the synthesis of acid phosphatase, which is abundant in osteoclast [14]. More recently, we have undertaken a more detailed investigation of the action of aluminum at the level of the bone cell [15]. Cultured mouse osteoblast-like (OB) and osteoclast-like (OC) cells were exposed to varying concentrations of $AlCl_3$. The trace element had a biphasic effect on tartrate-resistant acid phosphatase activity of OC cells and on alkaline phosphatase activity of OB cells: low aluminum concentrations ($\leq 1.5 \times 10^{-8}$ *M*) increased, whereas high aluminum concentrations inhibited both enzyme activities. Collagen (type I) synthesis of OB cells was not affected by aluminum concentrations from 10^{-11} to 10^{-7} *M*, whereas higher concentrations decreased ^{3}H-proline incorporation into collagenase-digestible proteins, indicating an inhibition of collagen synthesis. Both calcitriol and bovine (b) PTH (10^{-9} *M*) induced an inhibition of ^{3}H-proline incorporation under the same experimental conditions. Similarly, DNA synthesis (a marker of cell proliferation) and ornithine-decarboxylase activity (a marker of cell growth) of OB cells were increased at low, but decreased at high aluminum concentrations. Addition of calcitriol to the medium induced an inhibition of the stimulatory effect of low aluminum concentrations on ornithine-decarboxylase but its effect was not additive to the inhibitory action of high aluminum concentrations. The stimulatory effect of bPTH on this enzyme activity was inhibited by high doses of aluminum. Calcitriol and bPTH had no effect on DNA content of OB cells. The cAMP activity of OB or OC cells was not sensitive to even very high (10^{-5} *M*) aluminum concentrations. Finally, aluminum doses up to 10^{-6} *M* increased the release of ^{45}Ca and β-glucuronidase from mouse forelimb rudiments maintained in culture. Higher aluminum concentrations (10^{-5} *M*) inhibited the release of both, indicating reduced bone resorption.

In conclusion, these data show that aluminum influences several cellular functions in bone tissue. Its action is dose-dependent and may be biphasic in that lower concentrations of the trace element stimulate, whereas higher concentrations inhibit, several functions of osteoblasts and osteoclasts. These findings confirm the depressant effect of severe aluminum intoxication on bone matrix formation observed in vivo. Moreover, they demonstrate that this effect is not only due to the inhibition of collagen synthesis, but also to an interference with cell proliferation. Finally, our data show that aluminum directly influences bone resorption, and it interacts with the effects of PTH and calcitriol at the cellular level.

References

1 Plachot, J.J.; Cournot-Witmer, G.; Halpern, S.; Mendes, V.; Bourdeau, A.; Fritsch, J.; Bourdon, R.; Drüeke, T.; Galle, P.; Balsan, S.: Bone ultrastructure and x-ray microanalysis of aluminum-intoxicated hemodialyzed patients. Kidney int. *25:* 796 (1984).

2 Goodman, W.G.: Bone disease and aluminum: Pathogenic considerations. Am. J. Kidney Dis. *6:* 330 (1985).

3 Malluche, H.H.; Faugere, M.C.: Aluminum: toxin or innocent bystander in renal osteodystrophy. Am. J. Kidney Dis. *6:* 336 (1985).

4 Charhon, S.A.; Chavassieux, P.M.; Chapuy, M.C.; Boivin, G.Y.; Meunier, P.J.: Low rate of bone formation with or without histologic appearance of osteomalacia in patients with aluminum intoxication. J. Lab. clin. Med. *106:* 123 (1985).

5 Drüeke, T.; Cournot-Witmer, G.: Dialysis osteomalacia: clinical aspects and physiopathological mechanisms. Clin. Nephrol. *24:* S26 (1985).

6 Van de Vyver, F.L.; De Broe, M.E.: Aluminum in tissues. Clin. Nephrol. *24:* S37–S57 (1985).

7 Drüeke, T.; Lacour, B.; Touam, M.; Basile, C.; Bourdon, R.: Oral aluminum administration to uremic, hyperparathyroid, or vitamin D-supplemented rats. Nephron *39:* 10–17 (1985).

8 Burnatowska-Hledin, M.A.; Doyle, T.M.; Eadie, M.J.; Mayor, G.H.: 1,25-Dihydroxyvitamin D_3 increases serum and tissue accumulation of aluminum in rats. J. Lab. clin. Med. *108:* 96 (1986).

9 Demontis, R.; Leflon, A.; Fournier, A.; Tahiri, Y.; Hervé, M.; Morinière, P.; Abdull-Massih, Z.; Atik, H.; Belbrik, S.; Renaud, H.; Plaquet, R.: 1 Alpha (OH) vitamin D_3 increases plasma aluminum in hemodialyzed patients taking $Al(OH)_3$. Clin. Nephrol. *26:* 146 (1986).

10 Henry, H.L.; Norman, A.W.: Interactions between aluminum and the actions and metabolism of vitamin D_3 in the chick. Calcif. Tissue Int. *37:* 484 (1985).

11 Merke, J.; Lucas, P.A.; Szabó, A.; Helbing, F.; Hügel, U.; Drüeke, T.; Ritz, E.: $1,25(OH)_2D_3$ receptors and end-organ response in experimental aluminum intoxication. Kidney int. *32:* 204 (1987).

12 Talwar, H.S.; Hari Reddi, A.; Menczel, J.; Thomas, W.C.; Meyer, J.L.: Influence of aluminum on mineralization during matrix induced bone development. Kidney int. *29:* 1038 (1986).

13 Faugère, M.C.; Arnala, I.O.; Ritz, E.; Malluche, H.H.: Loss of bone resulting from accumulation of aluminum in bone of patients undergoing dialysis. J. Lab. clin. Med. *6:* 481 (1986).

14 Lieberherr, M.; Grosse, B.; Cournot-Witmer, G.; Thil, C.L.; Balsan, S.: In vitro effects of aluminum on bone phosphatases: a possible interaction with bPTH and vitamin D_3 metabolites. Calcif. Tissue Int. *34:* 280 (1982).

15 Lieberherr, M.; Grosse, N.B.; Cournot-Witmer, G.; Hermann-Erlee, M.P.M.; Balsan, S.: Aluminum action on mouse bone cell metabolism and response to PTH and $1,25(OH)_2D_3$. Kidney int. *31:* 736 (1987).

T. Drüeke, MD, INSERM U 90, Hôpital Necker-Enfants Malades,
161, rue de Sèvres, F-75743 Paris Cédex 15 (France)

Contr. Nephrol., vol. 64, pp. 113–119 (Karger, Basel 1988)

The Parathyroid Gland and Aluminum Overload: An Overview[1]

Jorge B. Cannata, J. Bernardino Díaz López, María J. Fernandez Menéndez, M. Jesus Virgós

Hospital General de Asturias, Facultad de Medicina de Oviedo, Spain

Knowledge of parathyroid hormone (PTH) and aluminum (Al) metabolism has increased in recent years. However, their relationship to each other still remains unclear.

In this overview, we shall focus the attention on the analysis of clinical and experimental evidences obtained during the last years, which have given information and support to better understand the Al effect on PTH function and the PTH influence on Al metabolism.

Al-PTH Relationship

First Evidences and Evolution of this Short Story

During the early 1970s, and as soon as Al was blamed for producing a new type of dialysis-related syndrome a relationship between Al and PTH was suspected. By then, the main goal of nephrologists in the management of renal osteodystrophy was to suppress medically the almost inevitable hyperparathyroidism found in chronic renal failure.

Orginally it was thought that the effect of Al on PTH secretion was indirect and only due to the changes observed in calcium and phosphate metabolism secondary to the use of aluminum hydroxide as a phosphate

[1] The aluminum-PTH project has received support from the Comisión Asesora de Investigación Científica y Técnica del Ministerio de Educación y Ciencia (CAICYT) 2837-C02 from Acción Integrada Hispano-Británica 21/72, Dirección de Política Científica, Spain, and from The Department of Nephrology-Hypertension, University of Antwerp, Belgium.

binder [1]. The possibility of gastrointestinal Al absorption and direct Al effect on PTH secretion did not have then solid support. Despite early warnings [2, 3], Al absorption from oral Al compounds was disregarded, being only recognized after a decade of controversy [4, 5].

At the end of the 1970s and beginning of the 1980s, the clinical and experimental works of Mayor et al. [6, 7] and Cann et al. [8] gave enough information to suspect a direct relationship between PTH and Al metabolism. Since then, two main questions have developed concerning the origin of successive hypotheses: (1) What is the role of PTH on Al metabolism? Is there any modulating effect of this hormone on Al distribution and further toxicity? (2) What are the mechanisms involved in the PTH dysfunction observed in Al overload?

If the two previous speculations are both valid: Which is the first? Which is the trigger? Do they interact? We shall try to give the possible answers to these questions in this article which could have important therapeutical implications.

Effect of PTH on Al Metabolism

Clinical Evidence

There is a great amount of evidence that PTH may play a *key role* in modifying the skeletal toxicity of Al. In the Denver experience, patients with overt hyperparathyroidism did not develop fracturing osteomalacia and did develop dialysis encephalopathy when dialyzed with high-Al dialysate [9]. In addition, Felsenfeld et al. [10] have demonstrated that subtotal parathyroidectomy could contribute to the development of osteomalacia in Al-intoxicated patients, Hodsman et al. [11] and Andress et al. [12] have also proved a high incidence of osteomalacia in hemodialysis patients lacking secondary hyperparathyroidism, suggesting a *protective* effect of PTH on bone Al accumulation and distribution in renal failure.

Experimental Experience

Mayor et al. [6, 7] demonstrated, in short-term Al intoxication in rats, that Al uptake and distribution in tissues might be a function of parathyroid activity. The latter could modulate Al gastrointestinal absorption and deposition. They also demonstrated, like Drüeke et al. [13], a different role of endogenous and exogenous hyperparathyroidism.

However, in other studies in uremic rats, Alfrey et al. [9, 14] did not find any effect of PTH on Al gastrointestinal absorption and excretion, but they did prove a PTH effect on the compartmentalization of Al, showing a decreased Al bone uptake after parathyroidectomy.

Aluminum Effect on Parathyroid Gland Function

Clinical Evidence

Nowadays it is well known that although the levels of PTH in serum are almost universally increased in patients with renal failure, a fair number of Al-intoxicated patients have 'inadequately' normal or low PTH levels and a great predisposition to develop hypercalcemia [11, 15]. If Al may diminish PTH secretion, it can thereby indirectly reduce the formation and mineralization of bone. Al could produce this *inadequate* hypoparathyroidism either directly by suppressing PTH formation, or secretion or indirectly through the elevations of serum calcium induced by the Al effect on bone.

Direct Mechanism

Suppressing PTH Formation or Secretion. This mechanism may have indirect support from works done by the groups of Coburn [16] and Llach [17] who found a poor PTH response to acute hypocalcemia in dialysis patients with proved Al toxicity. These studies did not investigate Al content of the parathyroid gland, but the results and correlations between bone and parathyroid gland content found by Cann et al. [8] and Verbueken et al. [18] allow us to suspect a high parathyroid Al content in these patients.

Unfortunately, in clinical studies it is difficult to define the *normal response* and what should be the *dialysis control group* as almost all non-Al-intoxicated dialysis patients have some degree of secondary hyperparathyroidism and therefore they belong to a *high-PTH baseline* responder group.

Inhibiting PTH Release (Ionic Al^{+3} Effect?). The effect of Al as an inhibitor of PTH release has been only proved under experimental conditions but so far there is no evidence to support it clinically. Recently, we demonstrated that the significant serum Al increase achieved by the use of desferrioxamine (DFO) was not able to depress acutely PTH release in patients with hyperparathyroidism undergoing hemodialysis [19, 20]. This lack of effect was probably due to the fact that Al-DFO as a complex might not be able to act on PTH release.

Indirect Mechanism

It is well known that Al-intoxicated patients have a great facility to develop hypercalcemia either spontaneously or when using calcium supplements or vitamin D metabolites. These facts incited us to search for proofs linking Al with calcium and PTH metabolism. In a group of 19 CAPD patients acute Al exposure depressed PTH release only in those patients who had a significant calcium increase almost parallel to the Al increase [21], suggesting that in some cases Al can also suppress PTH by inducing serum calcium elevations.

Experimental Experience

Inhibiting PTH Release (Al^{+3} Ionic Effect?). The direct, possibly ionic Al effect on PTH release has been proved in vitro by Morrissey and co-workers [22, 23], but this inhibiting effect has been achieved using an Al incubation medium with very low protein content and therefore with a significantly higher free Al^{+3} level (not protein-bound) than that currently found in renal patients.

In animal experiments, there are more conflicting results. Some authors like Robertson et al. [24] and Díaz López and De Broe et al. [personal communication] have found a decrease in PTH levels in Al-intoxicated rats. On the contrary, Goodman et al. [25, 26] failed to support this observation in dogs and rats. In a recent study we also did not find differences in PTH release (intact and rat PTH [INC]) between Al-intoxicated and control rats in both basal and post-stimulation values after acute hypocalcemia induced by EGTA [27]. However, all these experimental studies are quite difficult to compare due to the different assays used to assess the PTH response.

Closing Remarks

It is concluded that Al intoxication can influence PTH which in turn is able to modify the behavior of Al in the body [28, 29].

In short-term experimental studies of Al intoxication, PTH does not seem to be a major determinant. In contrast, in clinical experience, the existence of a long-standing low parathyroid function could be dangerous.

This fact clearly alerts one to the importance of removing parathyroid tissue in chronic renal failure in Al-intoxicated patients, as reduction in gland mass may clearly contribute to impaired bone regulation. On the other hand, severe hyperparathyroidism might be responsible for greater Al uptake and deposition.

Today, we have acquired at least enough information to support a hypothesis that links, in a comprehensive and simple way, this puzzling mutual Al-PTH relationship.

At the beginning, the common hyperparathyroidism present in chronic renal failure may be the main factor modulating Al uptake through different doorways of entry.

Al deposition progressively increases, so modifying the function of tissues such as parathyroid glands, bone, brain, etc.

This might explain the wide range of situations we face in clinical practice in Al-intoxicated patients, and the difficulties to fit each individual case in a preestablished scheme. This could also partly explain the great discrepancies observed between some clinical and experimental findings.

Although we still have to clear many questions, the bulk of the actual evidence attributes PTH with a key role in maintaining mineralization in Al-intoxicated uremic patients. This fact should always be kept in mind when deciding to treat secondary hyperparathyroidism.

References

1 Clarkson, E.M.; Luck, V.A.; Hynson, W.V.; Bailey, R.R.; Eastwood, J.B.; Woodhead, J.S.; Clements, V.R.; O'Riordan, J.L.; De Wardener, H.E.: The effect of aluminum hydroxide on calcium, phosphorus and aluminum balances, the serum parathyroid hormone concentration and the aluminum content of bone in patients with chronic renal failure. Clin. Sci. *4:* 519–531 (1972).

2 Berlyne, G.M.; Ben-Ari, J.; Pest, D.; Weinberger, J.; Stern, M.; Gilmore, G.R.; Levine, R.: Hyperaluminaemia from aluminium resins in renal failure. Lancet *ii:* 494–496 (1970).

3 Kaehny, W.D.; Hegg, A.P.; Alfrey, A.C.: Gastrointestinal absorption of aluminum from aluminum-containing antacids. New Engl. J. Med. *296:* 1389–1390 (1977).

4 Cannata, J.B.; Briggs, J.D.; Junor, B.J.R., Fell, G.S.: Aluminium hydroxide intake: real risk of aluminium toxicity. Br. med. J. [Clin. Res.] *286:* 1937–1938 (1983).

5 Andreoli, S.P.; Bergstein, J.M.; Sherrard, D.J.: Aluminum intoxication from aluminum-containing phosphate binders in children with azotemia not undergoing dialysis. New Engl. J. Med. *310:* 1079–1084 (1984).
6 Mayor, G.H.; Keiser, J.A.; Makdani, D.; Ku, P.K.: Aluminum absorption and distribution: effect of parathyroid hormone. Science *197:* 1187–1189 (1977).
7 Mayor, G.H.; Sprague, S.M.; Hourani, M.R.; Sanchez, T.V.: Parathyroid hormone-mediated aluminum deposition and egress in the rat. Kidney int. *17:* 40–44 (1980).
8 Cann, C.E.; Prussin, S.G.; Gordan, G.S.: Aluminum uptake by the parathyroid glands. J. clin. Endocr. Metab. *49:* 543–545 (1979).
9 Alfrey, A.C.: The case against aluminum affecting parathyroid function. Am. J. Kidney Dis. *6:* 309–312 (1985).
10 Felsenfeld, A.J.; Harrelson, J.M.; Gutman, R.A.; Wells, S.A.; Drezner, M.K.: Osteomalacia after parathyroidectomy in patients with uremia. Ann. intern. Med. *96:* 34–39 (1982).
11 Hodsman, A.B.; Sherrard, D.J.; Wong, E.G.C.; Brickman, A.S.; Lee, D.B.; Alfrey, A.C.; Singer, F.R.; Norman, A.W.; Coburn, J.W.: Vitamin-D-resistant osteomalacia in hemodialysis patients lacking secondary hyperparathyroidism. Ann. intern. Med. *94:* 629–637 (1981).
12 Andress, D.L.; Ott, S.M.; Maloney, N.A.; Sherrard, D.J.: Effect of parathyroidectomy on bone aluminum accumulation in chronic renal failure. New Engl. J. Med. *312:* 468–473 (1985).
13 Drüeke, T.; Lacour, B.; Touam, M.; Basile, C.; Bourdon, R.: Oral aluminum administration to uremic, hyperparathyroid, or vitamin-D-supplemented rats. Nephron *39:* 10–17 (1985).
14 Alfrey, A.C.; Sedman, A.; Chan, Y.L.: The compartmentalization and metabolism of aluminum in uremic rats. J. Lab. clin. Med. *105:* 227–233 (1985).
15 Cannata, J.B.; Briggs, J.D.; Junor, B.J.R.; Beastall, G.; Fell, G.S.: The influence of aluminium on parathyroid hormone levels in haemodialysis patients. Proc. Eur. Dial. Transpl. Ass. *19:* 244–247 (1983).
16 Kraut, J.A.; Shinaberger, J.H.; Singer, F.R.; Sherrard, D.J.; Saxton, J.; Miller, J.H.; Kurokawa, K.; Coburn, J.W.: Parathyroid gland responsiveness to acute hypocalcemia in dialysis osteomalacia. Kidney int. *23:* 725–730 (1983).
17 Andress, D.; Felsenfeld, A.J.; Voigts, A.; Llach, F.: Parathyroid hormone response to hypocalcemia in hemodialysis patients with osteomalacia. Kidney int. *24:* 364–370 (1983).
18 Verbueken, A.H.; Van de Vyver, F.L.; Visser, W.J.; Roels, F.; Van Grieken, R.E.; De Broe, M.E.: The use of laser microprobe mass analysis (LAMMA) for localizing multiple elements in soft and hard tissues. Biol. Trace Elem. *13:* 397–412 (1987).
19 Cannata, J.B.; De Francisco, A.M.; Amado, J.A.; Fernandez, M.D.; Junor, B.J.R.; Cuesta, M.V.; Ruiz, L.; Briggs, J.D.; Arias, M.: Parathyroid hormone release under the influence of aluminium-desferrioxamine complex; in Taylor, Aluminium and other trace elements in renal disease, pp. 38–44 (Ballière-Tindall, London 1986).
20 Cannata, J.B.; De Francisco, A.M.; Díaz de Greñu, M.C.; Lamas, J.; Arias, M.; Herrera, J.; Junor, B.J.: Hyperaluminaemia and parathyroid hormone release in haemodialysis patients. Abstr. Eur. Dial. Transpl. Ass. *1986:* 113.
21 Cannata, J.B.; Briggs, J.D.; Junor, B.J.R.; Fell, G.S.; Beastall, G.: Effect of acute

aluminium overload on calcium and parathyroid-hormone metabolism. Lancet *i:* 501–503 (1983).
22 Morrissey, J.; Rothstein, M.; Mayor, G.; Slatopolsky, E.: Suppression of parathyroid hormone secretion by aluminium. Kidney int. *23:* 699–704 (1983).
23 Morrissey, J.; Slatopolsky, E.: Effect of aluminum on parathyroid hormone secretion. Kidney int. *29:* suppl. 18, pp. S41–S44 (1986).
24 Robertson, J.A.; Felsenfeld, A.J.; Haygood, C.C.; Wilson, P.; Clarke, C.; Llach, F.: Animal model of aluminum-induced osteomalacia: role of chronic renal failure. Kidney int. *23:* 327–335 (1983).
25 Goodman, W.G.; Henry, D.A.; Horst, R.; Nudelman, R.K.; Alfrey, A.C.; Coburn, J.W.: Parenteral aluminum administration in the dog. Induction of osteomalacia and effect on vitamin D metabolism. Kidney int. *25:* 370–375 (1984).
26 Goodman, W.G.: Experimental aluminum-induced bone disease: studies in vivo. Kidney int. *29:* suppl. 18, pp. S32–S36 (1986).
27 Cannata, J.B.; Díaz de Greñu, M.C.; Allende, M.T.; De Francisco, A.M.; Herrera, J.; Junor, B.J.: Hyperaluminaemia and parathyroid hormone release: clinical and experimental studies (Abstract Spanish Society of Nephrology). Kidney int. (in press, 1987).
28 Cannata, J.B.; Briggs, J.D.; Junor, B.J.R.: in Drüeke, Rottembourg, Aluminium intoxication and parathyroid hormone function: clinical aspects. Aluminium et insuffisance rénale, pp. 227–238 (Gambro, Paris 1984).
29 Mayor, G.H.: The case for parathyroid hormone. Am. J. Kidney Dis. *6:* 306–308 (1985).

Jorge B. Cannata, MD, PhD, Unidad de Investigación, Hospital General de Asturias, Apartado 243, E-33080 Oviedo (España)

Contr. Nephrol., vol. 64, pp. 120–123 (Karger, Basel 1988)

The Histochemical Demonstration of Aluminum within Tissue Sections

A.J. Freemont

Department of Rheumatology, University of Manchester, Manchester, England

Histochemistry is a science in which chemical reactions are exploited to demonstrate the presence and distribution of specific chemical substances in a tissue. This necessitates adapting the techniques of pure chemistry to the requirements of microscopy and the complexities of the in situ biological medium.

The ultimate aim of the histochemist is to produce a final reaction product (FRP) that is specific, visible, insoluble, permanent and localized. Histochemistry can be adapted to any visualizing system from the naked eye to the electron microscope but the general principles are the same. A specimen (most frequently a 4- to 7-μm tissue section mounted on a microscope slide), is submitted to a series of chemical reactions. During these the substance under investigation is liberated from its original binding site in the tissue in a chemical state that will react with a dye. To permit localization the FRP must be produced immediately and precipitate as it forms, without diffusion, in sufficient quantities to be visualized. The reaction product must be not only insoluble in the staining solution but also in whatever other aqueous or organic solvent based media that are used in producing the finished preparation.

The histochemical localization of aluminum illustrates how difficult these general principles can be to apply. The most significant problem in demonstrating aluminum is liberating it from its binding sites. As it is not normally present in demonstrable amounts within human tissues differentiating endogenous from introduced aluminum is not a problem.

Although aluminum enters the tissues in various ways [1, 2], undoubtedly the most significant source is dialysis fluid from where it can rapidly gain access to all body compartments. The metal is neurotoxic, suppresses osteoblastic activity and hemopoiesis and probably causes parathyroid suppression [3]. That it induces abnormalities of cell function indicates that aluminum can bind with organic molecules; in addition it may be found sequestered within inorganic complexes. In general aluminum in man is probably found in 3 forms: readily reactive inorganic molecules, e.g. introduced aluminum salts; relatively inert inorganic substances, e.g. the hydroxyapatite crystal lattice of bone; stable complexes with large organic molecules – possibly in a form analogous to iron in hemosiderin. The ease with which aluminum can be liberated from these different sites varies considerably and explains some of the differences in perceived metal distribution using various staining regimes. The most widely used stain for aluminum is aluminon. The reactive component, aurine tricarboxylic acid (ACTA), forms colored compounds with a wide variety of metals but providing there is an excess of ammonium salts in the solution and the pH is rigorously controlled it is relatively specific for aluminum.

Aluminon localizes aluminum to bone surfaces and cement lines but is inhibited by an excess of phosphate ions [4], thus rarely, if ever, discloses aluminum elsewhere in bone.

The other commonly used dyes for staining aluminum belong to the solochrome family. One – solochrome azurine [5] – forms a relatively stable solution in buffer at pH 5, a pH low enough to decalcify bone. Decalcification liberates aluminum from the hydroxyapatite crystal lattice and this stain can therefore demonstrate the full extent of aluminum within bone. It is very sensitive. Unfortunately, it is not specific for aluminum reacting with a variety of rarely encountered metals such as beryllium or titanium. It also reacts with iron which may lead to errors of interpretation especially in dialysis patients receiving regular transfusions and/or parenteral iron therapy. In this laboratory we always stain a serial section with Perls' stain for iron as a control for the solochrome azurine technique.

Neither aluminon nor solochrome azurine convincingly or reproducibly demonstrates aluminum tightly bound to large organic molecules. In this context it is interesting to note that although solochrome azurine stains iron within bone it does not react with marrow iron even when the Perls' stain shows it to be present in abundance. Perls' reaction is performed at pH 2, sufficiently low to hydrolyze protein and liberate complexed metal. This type of treatment may be necessary to release aluminum from some of its complexes.

Solochrome azurine and ATCA are not stable at such low pH and cannot therefore be employed if one wishes to use acid hydolysis to identify aluminium in nonmineralized tissues.

Another solochrome salt-solochrome cyanine is stable at pH 2. This will stain aluminum within soft tissues but the color density of the FRP is so weak at this pH that staining is usually unconvincing and too unreliable for routine use. Other dyes will react with aluminum but our experience indicates them to be inferior to either solochrome azurine or aluminon.

A histochemical technique to be of practical use should be sensitive, specific and show the full extent of the substance being studied. Clearly none of the stains for aluminum meet all these criteria. Undoubtedly solochrome azurine at pH 5 is the most appropriate for demonstrating aluminum within bone but there is a need for a stain for aluminum in soft tissues. In this context we have recently had some encouraging results using a solochrome dye following microwave irradiation of the tissue and further studies are in hand to examine the potential of this technique.

At present histochemistry has an important role in the demonstration of aluminum in bone. It is reliable, quick and cheap. Unlike electron probe X-ray microanalysis and other physicochemical techniques, large areas of tissue can be readily examined. The solochrome azurine technique is also semiquantitative [6] and because the tissue is left intact is preferable to chemical analysis when distribution as well as amount are being studied. Because no stain reproducibly reacts with aluminum bound to complex organic molecules histochemical studies of tissues other than mineralized bone should be examined critically and greater credence given to the results of physicochemical and chemical analyses.

References

1 Pearse, A.G.E.: Histochemistry theoretical and applied; 3rd ed. (Churchill-Livingstone, Edinburgh 1972).

2 Williams, S.; Freemont, A.J.: Aerosol antiperspirants and axillary granulomata. Br. med. J. *288:* 1651–1652 (1984).

3 Freemont, A.J.; Day, J.P.; Ackrill, P.: Water treatment alone allows removal of aluminum from bone and reversal of the calcification defect in aluminum-related renal osteodystrophy. Abstr. 9th Int. Conf. of Nephrology, Int. Soc. of Nephrology, Los Angeles 1984, p. 48.

4 Vogel, A.J.: A text-book of macro and semi micro qualitative inorganic analysis; 4th ed. (Longmans, London 1954).
5 Denton, J.; Freemont, A.J.; Ackrill, P.; Day, J.P.: A new histochemical method for the demonstration of aluminium within bone; in Taylor, Aluminium and other trace elements in renal disease, part III, pp. 321–324 (Ballière, Tyndall, London 1986).
6 Denton, J.; Freemont, A.J.; Ball, J.: The detection and distribution of aluminium in bone. J. clin. Path. *37:* 136–142 (1984).

A.J. Freemont, MD, Department of Rheumatology, University of Manchester, Stopford Building, Oxford Road, Manchester M13 9PT (England)

Contr. Nephrol., vol. 64, pp. 124–133 (Karger, Basel 1988)

Localization of Aluminum in Tissues[1]

Armand H. Verbueken [a, 2], *Frank L. Van de Vyver* [b],
Etienne J. Nouwen [b], *René E. Van Grieken* [a], *Marc E. De Broe* [b]

Departments of [a]Chemistry and [b]Nephrology-Hypertension,
University of Antwerp (UIA), Wilrijk-Antwerp, Belgium

In clinical investigations, the micro-assay of trace elements is at present receiving great attention since abnormal elemental concentrations at the cellular level are likely to be the cause or consequence of disease. Such studies have become possible due to the development of sophisticated equipment for chemical microanalysis [1, 2]. Several methods have been applied in human pathology and cell biology to study a variety of biological tissues. They include electron probe X-ray microanalysis (EPXMA), secondary ion mass spectrometry or ion microscopy (SIMS), electron energy loss spectroscopy (EELS), nuclear microprobe analysis using charged particle induced X-ray emission (PIXE), and laser microprobe mass analysis (LAMMA) [3, 4].

The occurrence of aluminum-associated diseases is a pertinent example of clinical trace element toxicity. Bulk analysis of several organs of chronic dialysis patients reveals an elevated concentration of aluminum in bone, liver, spleen and parathyroid glands, while the brain accumulates less. Due to its larger mass, the total amount of aluminum in bone exceeds by far the quantities present in the other tissues.

A comparison of the aluminum levels in the different tissues examined in our laboratory [5–7] shows that the relation between aluminum accumulation and toxicity is complex. While bone and liver both contain high aluminum levels, toxicity has only been established in bone. Also, aluminum toxicity has been demonstrated in brain although brain alumi-

[1] This work was supported in part by the Belgian Ministry of Science Policy through Research Grants 80-85/10 and 82-87/47.

[2] A.H. Verbueken acknowledges the National Fund for Scientific Research (Belgium) for a position as Senior Research Assistant.

num levels are relatively low. This suggests that aluminum has accumulated at a critical ultrastructural site where it can interfere with essential biochemical processes before toxicity develops. Further elucidation of the mechanisms of aluminum-induced toxicity will, therefore, require the demonstration of biochemical interactions together with the detection of aluminum at the ultrastructural level.

Patients and Methods

In our studies microanalysis was applied to fresh tissue biopsies from chronic hemodialysis patients. Bone, liver, brain, thyroid and parathyroid glands have been studied using several techniques. Microanalytical procedures providing information on the chemical composition of microvolumes in histological sections (table I) are:

Electron Probe X-Ray Microanalysis. An electron beam is focused on a selected region of the specimen; the image of this area is visualized in the electron microscope. Characteristic X-rays are collected by crystal spectrometers or solid state detectors.

Laser Microprobe Mass Analysis. A high-power laser is focused on a histological section viewed by a light microscope (fig. 1). A selected area of a square micrometer of the tissue is evaporated; the created ions are separated in a time-of-flight mass spectrometer and detected. A complete mass spectrum covering all elements present in the microvolume is produced for each laser shot. Qualitatively, the advantages and limitations of LAMMA can be summarized as in table II. Considerable progress has been made lately to eliminate or reduce some of the drawbacks of LAMMA [8].

The moderate quality of the light microscopic image and the fact that usually unstained histological sections are used to avoid contamination make it necessary to rely on electron microscopy (EM) to correlate the ultrastructure of the specimen with the analytical results. EM micrographs have to be taken before and after LAMMA of the cellular regions of interest to provide records of the analyzed subcellular fragments and to measure the evaporated area. As an alternative, reflected light illumination may sometimes improve the visual differentiation of tissue structures. Incorporation of UV transillumination can give satisfactory images by enhancing the absorption contrast [9].

Secondary Ion Mass Spectrometry or Ion Microscopy. A beam of primary ions is focused on the surface of the histological sample. As a result of the ion bombardment, the atoms of the first atomic layers are sputtered away and ionized. These secondary ions reveal the atomic composition of the sample and produce a specific ion image. Selected images of the different ions can be displayed sequentially.

Electron Energy Loss Spectroscopy. A high-resolution electron energy spectrometer is combined with a conventional or scanning transmission electron microscope. Incident electrons are focused on a very small area of the section. After passing through the thin

Table I. Characteristics of the different microanalytical techniques

Analytical method	EPXMA	LAMMA	SIMS	EELS	Histochemistry
Excitation	electrons	photons	ions	electrons	complexing dye
Detection method	characteristic X-rays	ions	ions	electrons	light microscopy
Information	wavelength (WDS) energy (EDS)	mass-to-charge	mass-to-charge	energy loss	color
Lateral resolution	0.1–1 μm	1–2 μm (spot)	1 μm (ion image)	10 nm to 1 μm	few μm
Detection limits	WDS: (500 ppm) EDS: mg/g	ppm	ppm	(<100 ppm)	(>100 ppm)
Isotope detection	–	+	+	–	–
Molecular information	(–)	+	+	(+/–)	–
Morphological study	+ (EM)	+ (LM)	–	+ (EM)	+ (LM)
Sample thickness	0.1 μm few μm (SEM)	0.1–2 μm	⩾1 μm	⩽0.1 μm	5 μm
Destructive	(–)	+	+	–	(–)
Mapping	+	–	+	+	+
Quantification	+	(+)	(+)	+	(+)

WDS = Wavelength-dispersive spectrometer; LM = light microscopy; EDS = energy-dispersive spectrometer; EM = electron microscopy; SEM = scanning electron microscope.

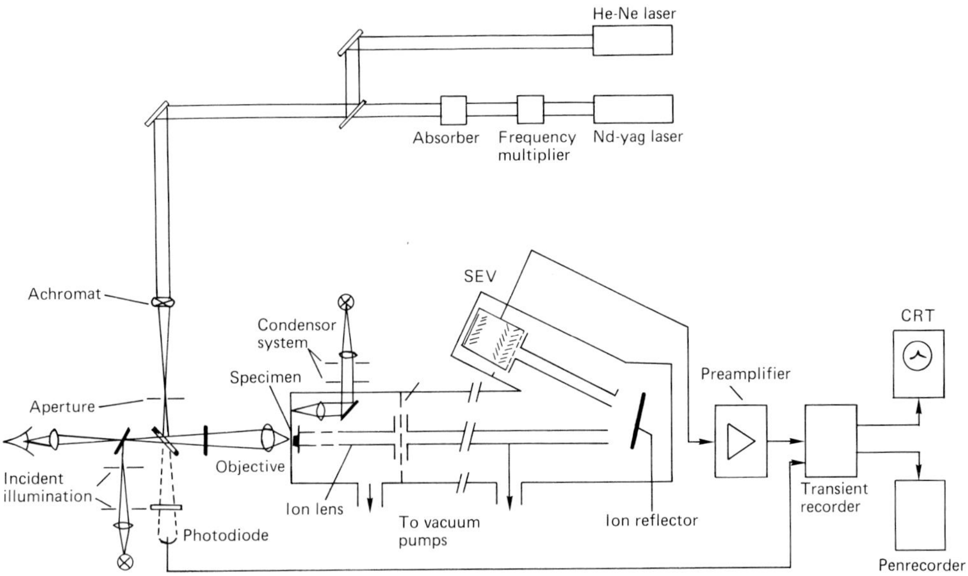

Fig. 1. Schematic representation of the laser microprobe mass analyzer (LAMMA-500).

Table II. Advantages and limitations of the LAMMA technique

Advantages	Limitations
Multi-element microanalysis (Li to U)	Destructive
Very low detection limits (down to ppm levels)	Spatial resolution may be insufficient for analysis of particular subcellular structures
Inorganic speciation	Quantification is tedious and still under investigation
Fingerprinting of organic molecules	Visualization of microstructures of biomedical samples is rather poor due to moderate quality of the light microscopic view
Isotopic information	Organic background interference at trace levels
Very fast analysis	Only thin films or microparticles can be analyzed (in the transmission geometry)
Spatial resolution around 1 µm	
Section thickness 0.1 (EM) to 2 µm (LM)	
Surface specific information can be obtained	
Positive and negative ion mode	

sample, the energy loss of the transmitted electrons due to interaction with the present atoms is measured.

Histochemical Staining (HS). The staining of tissues for aluminum usually involves the use of ammonium aurintricarboxylate (aluminon). Although they are routinely used, histochemical methods frequently lack sufficient specificity and suffer from interferences [10].

In our laboratory, we have used four of these methods for the localization of aluminum in human tissue biopsies, i.e. HS, EPXMA, SIMS, and LAMMA.

Results

In undecalcified bone sections of patients with aluminum-induced osteomalacia, aluminum is detected at the mineralized-bone boundaries (whether or not covered by osteoid), in histiocyte-like bone marrow cells, and occasionally at cement lines. The presence of aluminum in the bone was readily demonstrated by light microscopy and histochemistry using the Aluminon dye, which produced a red linear staining along the osteoid/calcified-bone boundaries. The localization of aluminum was verified and extended by using the instrumental LAMMA and SIMS techniques [11]. In patients with massive iron overload, LAMMA revealed the concomitant presence of iron and aluminum at the osteoid/calcified-bone boundaries (fig. 2) [12] and in bone marrow cells [13].

By combining LAMMA with EM, the localization of aluminum was established in fresh liver biopsies of dialysis patients [11, 12]. In these patients, who frequently demonstrated a concomitant iron overload, aluminum and iron were persistently detected together in electron-dense bodies within the liver cells, parenchymal as well as Kupffer cells (fig. 3). Histochemical staining for acid phosphatases verified the lysosomal nature of these subcellular structures.

In the thyroid and parathyroid glands from chronic dialysis patients, a lysosomal co-trapping of iron and aluminum was again observed. In the thyroid sample, aluminum and iron were concomitantly localized within lysosomes of follicular epithelial cells (fig. 4). Moreover, for the parathyroid glands we were able to demonstrate the presence of aluminum in secretory granules [14]. LAMMA spectra taken from groups of small granules within the cytoplasm near the nuclei revealed the presence of aluminum (fig. 5).

In cortical brain samples from dialysis patients with suspected dementia, small amounts of aluminum were detected in electron-dense

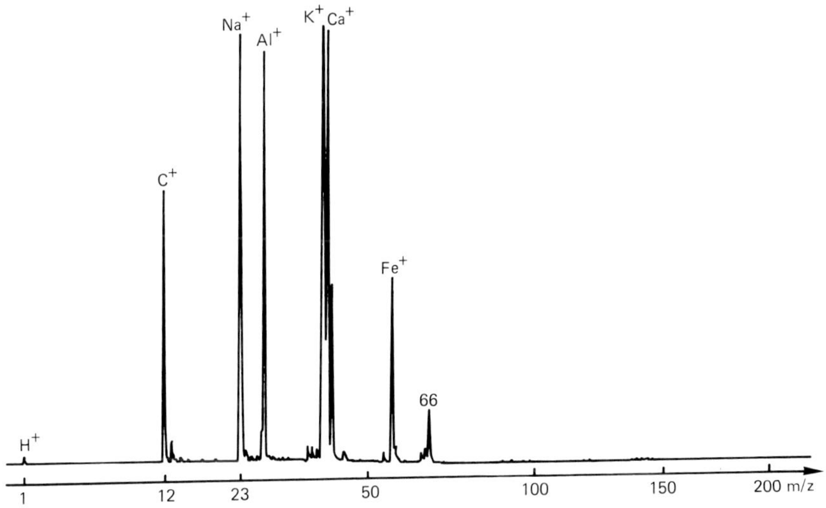

Fig. 2. LAMMA spectrum taken from a bone section of a patient with iron and aluminum overload and dialysis osteomalacia. The analyzed region corresponds to the osteoid/calcified-bone boundary. The concurrent presence of aluminum and iron is demonstrated.

cytoplasmic bodies (lysosomes, lipofuscin granules). The very low concentration of aluminum in the brain, however, precluded any further subcellular localization of this toxic element. The localization of aluminum in the different tissues of chronic dialysis patients, as determined in our laboratory, is summarized in table III.

Discussion

The ultrastructural localization of aluminum bears an important relationship with toxicity. The sequestration of aluminum in lysosomes of hepatocytes and Kupffer cells might explain the absence of proven toxicity of aluminum for the liver. Lysosomal storage of aluminum can be considered an important mechanism for neutralizing the metal's deleterious effects. In contrast, the presence of aluminum at the extracellular osteoid/calcified-bone boundary possibly causes the bone mineralization defect seen in aluminum intoxication. Similarly, the occurrence of

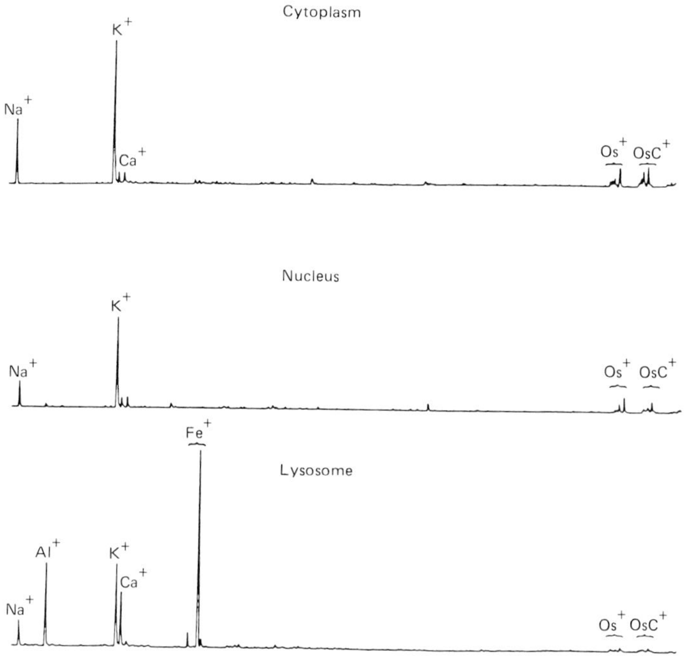

Fig. 3. LAMMA spectra obtained from different subcellular sites of a liver section from a dialysis patient with osteomalacia. Significant iron and aluminum signals are seen at the level of the hepatocyte lysosomes.

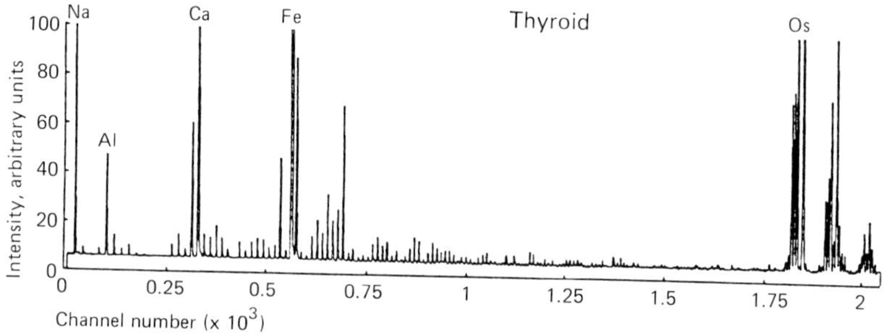

Fig. 4. LAMMA spectrum of a thyroid lysosome (follicular epithelial cell). A concomitant localization of aluminum and iron is demonstrated. The osmium signals result from postfixation of the tissue.

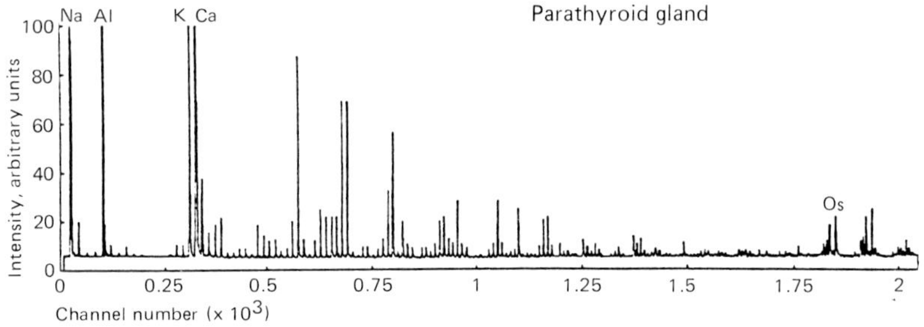

a

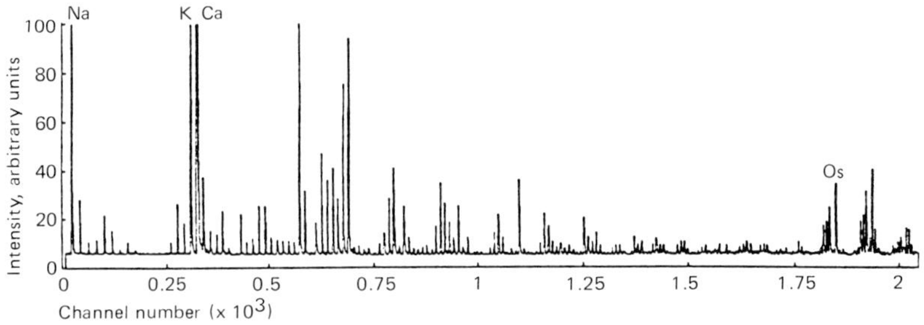

b

Fig. 5. LAMMA spectra of parathyroid gland from a chronic dialysis patient. *a* Taken from small granules within the cytoplasm (presumably secretory granules) demonstrates the presence of aluminum. *b* Taken from a nearby region without any visible granules does not show aluminum. The osmium signals result from the postfixation procedure.

Table III. Localization of aluminum in tissues of chronic dialysis patients

Bone	osteoid/calcified-bone boundary, marrow cells
Liver	lysosomes (hepatocytes and Kupffer cells)
Brain	lipofuscin granules, lysosomes
Parathyroids	lysosomes, secretory granules
Thyroid gland	lysosomes

aluminum in secretory granules of parathyroid cells may inhibit parathyroid hormone (PTH) release resulting in the relatively low serum PTH levels found in dialysis patients with aluminum-induced osteomalacia.

LAMMA appears to be the method of choice to determine the ultrastructural localization of aluminum and to define its occurrence with respect to other elements such as iron. LAMMA was preferred in view of its high sensitivity for simultaneous multiple element detection combined with a spatial resolution of analysis of about 1 μm. Using LAMMA, we have often found aluminum and iron at the same ultrastructural sites, e.g. in liver lysosomes, thyroid and parathyroid gland lysosomes, at the osteoid/calcified-bone boundaries, in bone marrow cells with histiocytic appearance.

The ultrastructural localization of aluminum in histological sections by means of microanalytical techniques can reveal the presence of this element and of associated elements at a particular biochemically critical site, where cellular toxicity may be induced. In this way, the mechanisms of aluminum-induced toxicity can be better understood.

Methodological improvements are expected. Sophisticated equipment, e.g. EELS imaging, combined with refined sample preparation techniques (cryofixation) will allow the determination of ever decreasing elemental concentrations with a more refined spatial resolution.

References

1 Baker, D.; Kupke, K.G.; Ingram, P.; Roggli, V.L.; Shelburne, J.D.: Microprobe analysis in human pathology. Scann. electron Microsc. *11:* 659–680 (1985).

2 Verbueken, A.H.; Van de Vyver, F.L.; Van Grieken, R.E.; De Broe, M.E.: Microanalysis in biology and medicine: ultrastructural localization of aluminum. Clin. Nephrol. *24:* S58–S77 (1985).

3 Verbueken, A.H.; Bruynseels, F.J.; Van Grieken, R.E.: Laser microprobe mass analysis: a review of applications in the life sciences. Biomed. Mass Spectrom. *12:* 438–463 (1985).

4 Verbueken, A.H.; Van de Vyver, F.; De Broe, M.; Van Grieken, R.: Applications of laser microprobe mass analysis in medicine. CRC crit. Rev. Clin. Lab. Sci. *24:* 263–285 (1987).

5 Van de Vyver, F.L.; De Broe, M.E.: Aluminum in tissues. Clin. Nephrol. *24:* S37–S57 (1985).

6 De Broe, M.E.; Van de Vyver, F.L.; Bekaert, A.B.; D'Haese, P.; Paulus, G.J.; Visser, W.J.; Van Grieken, R.; de Wolff, F.A.; Verbueken, A.H.: Correlation of serum aluminum values with tissue aluminum concentration. Contr. Nephrol., vol. 38, pp. 37–46 (Karger, Basel 1984).

7 De Broe, M.E.; Van de Vyver, F.L.; E Silva, F.J.; D'Haese, P.C.; Verbueken, A.H.: Measuring aluminum in serum and tissues: overview and perspectives. Nefrologia *VI:* S41–S46 (1986).

8 Proc. 3rd Int. Laser Microprobe Mass Spectrometry Workshop, 1986 Antwerp.

9 Kaufmann, R.: Laser microprobe mass analysis (LAMMA): current state of the art with special emphasis on bio-medical applications; in Heinrich, Microbeam analysis – 1982, pp. 341–358 (San Francisco Press, San Francisco 1982).

10 Verbueken, A.H.; Van de Vyver, F.L.; Visser, W.J.; Van Grieken, R.E.; De Broe, M.E.: Laser microprobe mass analysis (LAMMA) to verify the aluminon staining of bone. Stain Technol. *61:* 287–295 (1986).
11 Verbueken, A.H.; Van de Vyver, F.L.; Van Grieken, R.E.; Paulus, G.J.; Visser, W.J.; D'Haese, P.; De Broe, M.E.: Ultrastructural localization of aluminum in patients with dialysis-associated osteomalacia. Clin. Chem. *30:* 763–768 (1984).
12 Van de Vyver, F.L.; Vanheule, A.O.; Verbueken, A.H.; D'Haese, P.; Visser, W.J.; Bekaert, A.B.; Van Grieken, R.E.; Buyssens, N.; De Keersmaecker, W.; Van den Bogaert, W.; De Broe, M.E.: Patterns of iron storage in patients with severe renal failure. Contr. Nephrol., vol. 38, pp. 153–166 (Karger, Basel 1984).
13 Van de Vyver, F.L.; Verbueken, A.H.; Visser, W.J.; Van Grieken, R.E.; De Broe, M.E.: Localisation of aluminium and iron by histochemical and laser microprobe mass analytical techniques in bone marrow cells of chronic haemodialysis patients. J. clin. Path. *37:* 837–838 (1984).
14 Verbueken, A.H.; Van de Vyver, F.L.; Nouwen, E.J.; Roels, F.; De Broe, M.E.; Van Grieken, R.E.: Laser microprobe mass analysis (LAMMA) of parathyroid glands from dialysis patients; in Brätter, Schramel, Trace element analytical chemistry in medicine and biology, vol. 4, pp. 443–450 (de Gruyter, Berlin 1987).

R.E. Van Grieken, PhD, Department of Chemistry, University of Antwerp (UIA), B-2610 Wilrijk-Antwerp (Belgium)

Contr. Nephrol., vol. 64, pp. 134–143 (Karger, Basel 1988)

Iron Overload and Bone Disease in Chronic Dialysis Patients

Frank L. Van de Vyver [a], *Walter J. Visser* [b], *Patrick C. D'Haese* [a], *Marc E. De Broe* [a]

[a]Department of Nephrology-Hypertension, University of Antwerp (UIA), Antwerp, Belgium; [b]Clinical Research Group for Bone Metabolism, University Hospital of Utrecht, Utrecht, The Netherlands

We were the first to observe that in dialysis patients with aluminum and iron accumulation both elements may be found concomitantly at the calcified bone boundary (CBB)[1] [1]. To demonstrate this, laser microprobe mass analysis (LAMMA) was used, a method which is not subject to chemical interferences [2]. This technique has proven to be a useful tool to verify histochemical staining procedures of trace elements [3]. Later, it was observed that in the presence of both aluminum and iron, the Aluminon® (BDH, Poole, England) staining, which is used for the histochemical localization of aluminum, may yield aberrant results [4]. In fact, some chronic dialysis patients with osteomalacia show purplish lines at the CBB instead of the bright-red lines commonly observed in aluminum-induced osteomalacia. LAMMA identified both aluminum and iron at the purplish lines [4]. It remained unclear, however, whether iron by itself in the absence of aluminum gives purplish lines at the CBB in the Aluminon staining. Furthermore, the role of iron in the development of bone disease in patients with chronic renal failure has not been documented, although it has been suggested to play a role in the osteopenia seen in thalassemic patients with heavy iron overload [5], and in the osteoporosis found in Bantu siderosis [6].

[1] The CBB is the surface of the mineralized part of the bone trabeculae whether or not this is covered with unmineralized osteoid. The bone disease is evaluated in the cancellous part of a complete transiliac bone biopsy.

In order to answer these questions we studied 120 bone biopsy specimens which were obtained for clinical purposes in chronically dialyzed patients. A standard procedure was used to obtain the biopsies [7]. One specimen was used for histology, the other for chemical measurements of iron and aluminum. The chemical determinations were not performed in all patients.

Bone Chemical Analysis. Both iron and aluminum were measured using electrothermal atomic absorption spectrometry (ETAAS). The methods have been described more extensively elsewhere [8, 9].

Bone Histology. Histologic and histomorphometric investigation of the second bone biopsy sample was performed after 24 h fixation in Burkhardt's solution and subsequent transfer to absolute methanol [7]. Undecalcified, 4 μm-thick bone sections were stained according to Goldner for descriptive histology. Using tri-ammonium aurin tricarboxylate (Aluminon), aluminum deposits were made visible [10]. Perls' staining was used for the detection of iron [11].

In all cases bone histomorphometry was performed by means of a grid consisting of successive semicircles [12] on bone sections stained with Goldner's, Aluminon and Perls' stainings. The following histomorphometric parameters were obtained in cancellous bone [13]: volumetric density of osteoid (Vvos%); osteoblast surface percentage (ob%); osteoid surface percentage (os%); ROBA% (relative osteoblastic activity) [14]; Aluminon-positive surface percentage (Al%); Perls-positive surface percentage (Fe%).

The classification of bone disease was based on that of other authors [15]. However, tetracycline labelling was not performed, since many patients were refered from other centers. Also, the measurement of mineralization rate with this procedure may show variability from site to site. As proposed by Merz and Schenk, ROBA% was used to evaluate osteoblastic activity [14]. In view of these considerations, bone lesions were classified as follows (table I). The bone was considered to be normal when the osteoid volume was not increased, fibrosis was absent and ROBA% was normal. The existence of osteopenia or osteosclerosis was not considered. In order to study the relation between iron, aluminum, and bone disease, only patients with positive Perls and/or Aluminon staining were considered further. In this way 40 patients were selected.

The composition of the studied patient group is given in table II. In view of the selection, every patient showed positivity in at least one of the 2 histochemical stainings. As criterion for positivity only the CBB was considered and the staining was said to be positive when the appropriate parameter (Al% for the Aluminon stain and Fe% for the iron stain) was 1% or more. In the Perls' staining these lines were always blue, but the color intensity was variable. Positive Perls' stainings were ascribed to the presence of iron at the CBB. In the Aluminon staining the color of the lines varied from bright-red to purplish.

In an earlier paper, we have demonstrated that purplish staining is commonly due to the concomitant presence of aluminum and iron at the

Table I. Classification of bone disease in severe chronic renal failure

	Vvos%	ROBA%	fibrosis
Normal bone	N or ↙	N	–
Early hyperparathyroidism	↙ or N or ↗	↗	–
Osteitis fibrosa	N or ↙	N or ↗	+
Mixed disease	↗	↙ or N or ↗	+
Osteomalacia	↗	N or ↙	–
Aplastic bone disease	N or ↙	↙	– or +

Table II. Composition of the studied patient group (n = 40)

Histochemical stainings at the CBB		Number of patients			
Aluminon	Perls'	total	bone aluminum below 20 μg/g	bone aluminum level of 20 μg/g or more	no bone aluminum level available
+	–	17	1	15	1
–	+	6	3	2	1
+	+	17	9	4	4
All patients (positive Aluminon and/or positive Perls' stainings)		40	13	21	6

CBB [4]. The question remained, however, whether the sole presence of iron could be responsible for the occurrence of purplish lines at Aluminon staining. To answer this question, the patients with positive Fe% were studied. Many of them showed bone aluminum concentrations below 20 μg/g (fig. 1). In patients with end-stage renal failure such aluminum levels are considered to be low or moderately low [16]. Furthermore, the occurrence of a positive Aluminon staining in a patient with bone aluminum below 20 μg/g and a negative Perls' staining is highly exceptional [17]. On the contrary, many of our patients with low bone aluminum and positive iron stainings showed positive Aluminon stainings (fig. 1). In these cases, the lines at the CBB were purplish and in some of them the

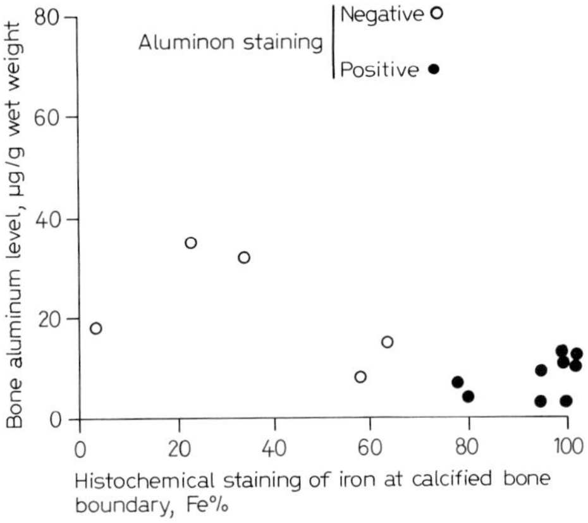

Fig. 1. Bone aluminum levels (µg aluminum per gram of wet transiliac bone) in cases where Perls' staining is positive at the CBB. In all patients with extensive blue lines at the CBB with Perls' staining, the Aluminon staining shows purple lines at this location, even when bone aluminum levels were relatively low.

bone aluminum levels are exceptionally low (below 5 µg/g in three). Therefore, it was assumed that in these cases the positive Aluminon staining could not be attributed to the presence of aluminum and the lines were ascribed to the sole presence of iron. An additional argument is the observation that Fe% was 78 or more in all the Aluminon-positive, low-bone-aluminum patients (fig. 1). On the other hand, Aluminon appears to be less sensitive to iron than Perls' stain, since a number of patients are Perls-positive and Aluminon negative at the CBB.

The present study confirms the earlier observation that Aluminon staining is not always positive when bone aluminum levels are increased [17]. This is thought to be due to the variable distribution of aluminum: it has been suggested that in hyperparathyroid bone disease aluminum is more evenly distributed throughout the bone trabeculae than in osteomalacia. Consequently, local aluminum levels may be lower than the detection limit for the Aluminon staining and/or the aluminum inside the bone mineral may be out of the reach of the stain [18].

In the present study a similar comparison was made between iron histochemical results and bone iron levels (fig. 2). As expected, the discrepancy between the results of iron staining at the CBB and bone iron measurement is even more pronounced than in the case of aluminum [17].

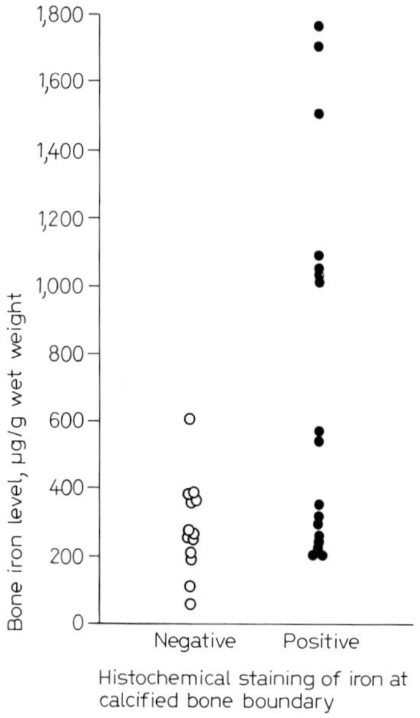

Fig. 2. Bone iron levels (µg iron per gram of wet transiliac bone) in patients with negative (Fe% below 1) and with positive Perls' staining at the CBB (Fe% of 1 or more).

Although in most patients with severe bone iron overload Fe% is positive, the results of iron staining are variable in those with bone iron between 200 and 400 µg/g. This is not surprising, since most iron contained in cancellous bone is found in bone marrow macrophages.

Bone disease in chronic dialysis patients is influenced by many factors, impeding the clinical study of each of them separately. However, the comparison of bone histomorphometry with biochemical data (e.g. parathyroid status, bone aluminum load, etc.) may suggest some relationships. In the present study, bone disease of two groups of patients was compared (fig. 3). The first group consisted of patients with iron accumulation at the CBB (Fe% positive) and low or moderate bone aluminum levels (less than 20 µg aluminum per gram of wet bone). The second group consisted of patients with considerable bone aluminum accumulation (more than 20 µg aluminum per gram of wet bone) and positive Aluminon staining. The numerous factors influencing bone metabolism in dialysis patients [19] impede the establishment of an appropriate control group. However, only bone iron and aluminum

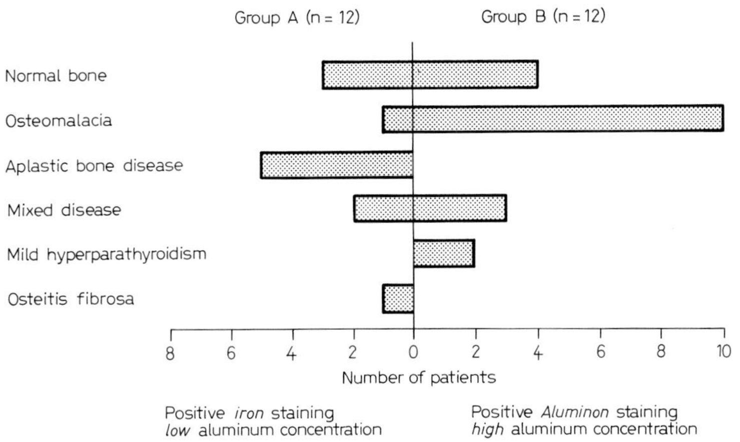

Fig. 3. Incidence of different bone diseases in two groups of chronic dialysis patients. Group A (n=12) consisted of patients who were Fe%-positive and had bone aluminum levels below 20 μg/g. Group B (n=19) included all patients who were Al%-positive and had bone aluminum levels of 20 μg/g or more. Diagnosis of bone disease was based on bone histology according to table I.

accumulation were used as criteria to classify patients in groups A and B, respectively. Therefore, it was felt that the increased stores of one of the two elements was the major determinant of the eventual differences in bone histology. As seen in figure 3, osteomalacia is frequent in the second group, which is in agreement with the established relationship between bone aluminum accumulation and the presence of osteomalacia [20]. In the first group the most frequent bone lesion is aplastic bone disease. In end-stage renal failure this histologic picture is relatively rare and has been attributed to aluminum toxicity [21]. The rarity of this disorder of bone formation in dialysis patients without prominent aluminum accumulation [21] and its high frequency in patients with stainable iron at the CBB suggest an association between bone iron accumulation and adynamic bone disease.

Impaired bone formation has also been observed in other conditions characterized by iron overload. In hereditary hemochromatosis (HH), osteoporosis is a frequent finding [22]. However, the osteoporosis has been attributed to hypogonadism [23] and not to iron. Patients with HH may show hyperparathyroidism [24]. In other cases of HH iron accumulation in the parathyroid glands causes hypoparathyroidism [25]. Like in our iron-loaded patients, bone histomorphometry in HH demonstrates that osteoid seam thickness is often decreased [23]. Osteoporosis is also a prominent feature of Bantu siderosis [6]. Both ascorbic acid deficiency as well as direct toxicity of iron have been offered as explanations [26]. In thalassemia major with secondary hemochromatosis, impaired bone mineralization and apposition rate associated with iron accumulation in bone have been observed [5]. Again, clinical scurvy has been implicated in this condition [27].

The existence of an association between iron overload and decreased osteoblastic activity does not necessarily imply a causal relationship, but a study of bone remodelling in pigs supports this contention [28]. The patient in chronic dialysis (especially in the case of hemodialysis) is particularly prone to develop iron and/or aluminum accumulation [29, 30]. Anemia is common and sometimes requires transfusion especially in anephric patients. Blood transfusions are also given as part of pre-transplant programs.

The recognition that iron and aluminum behave similarly in many situations has proved to be a fruitful concept [1]. The concomitant accumulation of aluminum and iron in chronic dialysis patients is not restricted to the CBB, but has also been observed in bone marrow macrophages [31] and in lysosomes of hepatocytes [32], Kupffer cells [32], and parathyroid chief cells [33]. The present study shows, however, that bone pathology in end-stage renal disease is different in cases of iron or aluminum accumulation. Repetition of earlier animal experiments [34], using iron chloride instead of aluminum chloride also has not resulted in the development of osteomalacia [35].

The hypothesis of adynamic bone disease due to iron-induced hypoparathyroidism is supported by the observation of hypoparathyroidism in other conditions associated with iron overload [25] and suppression of PTH secretion by iron in vitro [36]. Furthermore, experiments in rats suggest that parathyroidectomy is essential in the development of aplastic bone disease in aluminum-loaded animals with renal failure [37]. On the contrary, the experiments in iron-loaded pigs with normal renal function did not show decreased plasma parathormone levels [38]. Finally, the role of liver disease in iron-loaded dialysis patients deserves consideration.

The present study emphasizes the need to differentiate bone iron and aluminum accumulation from each other. Bone histology and aluminum histochemistry are insufficient to make this distinction [4]. Therefore, the staining for iron of every bone biopsy specimen belonging to a patient with severe renal failure is essential. In addition, bulk analysis with severe quality control [9] is mostly required. Microanalytical studies may provide additional insight in selected cases [32]. Although animal studies have demonstrated that aluminum may cause adynamic bone disease without osteomalacia [38], it remains a question how much iron had been accumulated in the bones of dialysis paticnts with this diagnosis [21].

To our knowledge this is the first report of an association between iron accumulation in end-stage renal disease and the presence of a specific bone disease. Further studies are required to demonstrate the etiologic role of iron in the reduction of bone formation in these patients.

References

1 Van de Vyver, F.L.; Vanheule, A.O.; Verbueken, A.H.; D'Haese, P.; Visser, W.J.; Bekaert, A.B.; Van Grieken, R.E.; Buyssens, N.; De Keersmaecker, W.; Van den Bogaert, W.; De Broe, M.E.: Patterns of iron storage in patients with severe renal failure. Contr. Nephrol. vol. 38, pp. 153–166 (Karger, Basel 1984).

2 Verbueken, A.H.; Van de Vyver, F.L.; Van Grieken, R.E.; De Broe, M.E.: Microanalysis in biology and medicine: ultrastructural localization of aluminum. Clin. Nephrol. *24:* S58–S77 (1985).

3 Van de Vyver, F.L.; Verbueken, A.H.; Van Grieken, R.E.; De Broe, M.E.: Laser microprobe mass analysis: a tool for evaluating histochemical staining of trace elements. Clin. Chem. *31:* 351–352 (1985).

4 Verbueken, A.H.; Van de Vyver, F.L.; Visser, W.J.; Van Grieken, R.E.; De Broe, M.E.: Laser microprobe mass analysis (LAMMA) to verify the Aluminon staining of bone. Stain Technol. *61:* 287–294 (1986).

5 Vernejoul, M.C. de; Girot, R.; Gueris, J.; Cancela, L.; Bang, S.; Bielakoff, J.; Mautalen, C.; Goldberg, D.; Miravet, L.: Calcium phosphate metabolism and bone disease in patients with homozygous thalassemia. J. clin. Endocr. Metab. *54:* 276–281 (1982).

6 Seftel, H.C.; Malkin, C.; Schmaman, A.; Abrahams, C.; Lynch, S.R.; Charlton, R.W.; Bothwell, T.H.: Osteoporosis, scurvy and siderosis in Johannesburg Bantu. Br. med. J. *i:* 642–646 (1966).

7 Visser, W.J.; Van de Vyver, F.L.: Aluminium-induced osteomalacia in severe chronic renal failure (SCRF). Clin. Nephrol. *24:* S30–S36 (1985).

8 Van de Vyver, F.L.; Vanheule, A.O.; Majelyne, W.M.; D'Haese, P.; Blockx, P.; Bekaert, A.B.; Buyssens, N.; De Keersmaecker, W.; De Broe, M.E.: Serum ferritin as a guide for iron stores in chronic hemodialysis patients. Kidney int. *16:* 451–458 (1984).

9 D'Haese, P.; Van de Vyver, F.L.; Wolff, F.A. de; De Broe, M.E.: Measurement of aluminum in serum, blood, urine and tissues of chronic hemodialyzed patients by use of electrothermal atomic absorption spectrometry. Clin. Chem. *31:* 24–29 (1985).

10 Irwin, D.A.: The demonstration of aluminum in animal tissues. Archs. industr. Health *12:* 218–220 (1955).

11 Perls, M.: Nachweis von Eisenoxyd in gewissen Pigmenten. Virchows Arch. *39:* 42–48 (1867).

12 Merz, W.A.: Die Streckenmessung an gerichteten Strukturen in Mikroskop und ihre Anwendung zur Bestimmung von Oberflächen-Volumen-Relationen im Knochengewebe. Mikroskopie *22:* 132–142 (1967).

13 Duursma, S.A.; Visser, W.J.; Njio, L.: A quantitative histological study of bone in 30 patients with renal insufficiency. Calcif. Tissue Res. *9:* 216–225 (1972).

14 Merz, W.A.; Schenk, R.K.: A quantitative histological study on bone formation in human cancellous bone. Acta anat. *76:* 1–15 (1970).

15 Ott, S.M.; Maloney, N.A.; Coburn, J.W.; Alfrey, A.C.; Sherrard, D.J.: The prevalence of bone aluminum deposition in renal osteodystrophy and its relation of the response to calcitriol therapy. New Engl. J. Med. *307:* 709–713 (1982).

16 De Broe, M.E.; Van de Vyver, F.L.; Bekaert, A.B.; D'Haese, P.; Paulus, G.J.; Visser, W.J.; Van Grieken, R.E.; Wolff, F.A. de; Verbueken, A.H.: The correlation of serum aluminium values with tissue aluminium concentration. Contr. Nephrol., vol. 38, pp. 37–46 (Karger, Basel 1984).

17 Visser, W.J.; Van de Vyver, F.L.; Verbueken, A.H.; D'Haese, P.C.; Bekaert, A.B.; Van Grieken, R.E.; Duursma, S.A.; De Broe, M.E.: Evaluation of different techniques used to determine aluminium in patients with chronic renal failure; in Priest, Metals in bone, pp. 433–441 (MTP Press, Lancaster 1985).

18 Denton, J.; Freemont, A.J.; Ball, J.: Detection and distribution of aluminum in bone. J. clin. Path. *37:* 136–142 (1984).

19 Sherrard, D.J.: Aluminum and renal osteodystrophy. Semin. Nephrol. *6:* suppl. 1, pp. 5–11 (1986).

20 Platts, M.M.; Goode, G.C.; Hislop, J.S.: Composition of the domestic water supply and the incidence of fractures and encephalopathy in patients on home dialysis. Br. med. J. *ii:* 657–660 (1977).

21 Llach, F.; Felsenfeld, A.J.; Coleman, M.D.; Keveney, J.J.; Pederson, J.A.; Medlock, T.R.: The natural course of dialysis osteomalacia. Kidney int. *29:* suppl. 18, pp. S74–S79 (1986).

22 Sèze, S. de; Solnica, J.; Mitrovic, D.; Miravet, L.; Dorfmann, H.: Joint and bone disorders and hypoparathyroidism in hemochromatosis. Semin. Arthrit. Rheumatism *2:* 71–94 (1972).

23 Pawlotsky, Y.; Lancien, Y.; Roudier, G.; Hany, Y.; Louboutin, J.Y.; Ferrand, B.; Bourel, M.: Histomorphométrie osseuse et manifestations ostéo-articulaires de l'hémochromatose idiopathique. Revue Rhum. *46:* 91–99 (1979).

24 Pawlotsky, Y.; Simon, M.; Hany, Y.; Brissot, P.; Bourel, M.: High plasma parathyroid hormone levels and osteo-articular changes in primary haemochromatosis. Scand. J. Rheum. *4:* suppl., p. 8 (1975).

25 Sherman, L.A.; Pfeferbaum, A.; Brown, E.B.: Hypoparathyroidism in a patient with long standing iron storage disease. Ann. intern. Med. *73:* 259–261 (1970).

26 Lynch, S.R.; Seftel, H.C.; Wapnick, A.A.; Charlton, R.W.; Bothwell, T.H.: Some aspects of calcium metabolism in normal and osteoporotic Bantu subjects with special reference to the effects of iron overload and ascorbic acid depletion. S. Afr. J. med. Sci. *35:* 45–56 (1970).
27 Cohen, A.; Cohen, I.J.; Schwartz, E.: Scurvy and altered iron stores in thalassemia major. New Engl. J. Med. *304:* 158–160 (1981).
28 Vernejoul, M.C. de; Pointillart, A.; Cywiner Golenzer, C.; Morieux, C.; Bielakoff, J.; Modrowski, D.; Miravet, L.: Effects of iron overload on bone remodelling in pigs. Am. J. Path. *116:* 377–384 (1984).
29 Gokal, R.; Millard, P.R.; Weatherall, D.J.; Callender, S.T.; Ledingham, J.G.; Olliver, D.A.: Iron metabolism in hemodialysis patients. Q. Jl. Med. *191:* 369–391 (1979).
30 Rahman, H.; Channon, S.M.; Parkinson, I.S.; Skillen, A.W.; Ward, M.K.; Kerr, D.N.S.: Aluminum in the dialysis fluid. Clin. Nephol. *24:* suppl. 1, pp. S78–S83 (1985).
31 Van de Vyver, F.L.; Verbueken, A.H.; Visser, W.J.; Van Grieken, R.E.; De Broe, M.E.: Localisation of aluminium and iron by histochemical and laser microprobe mass analytical techniques in bone marrow cells of chronic hemodialysis patients. J. clin. Path. *37:* 837 (1984).
32 Verbueken, A.H.; Van de Vyver, F.L.; Van Grieken, R.E.; Paulus, G.J.; Visser, W.J.; D'Haese, P.; De Broe, M.E.: Ultrastructural localization of aluminum in patients with dialysis-associated osteomalacia. Clin. Chem. *30:* 763–768 (1984).
33 Verbueken, A.H.; Van de Vyver, F.L.; Nouwen, E.J.; Roels, F.; De Broe, M.E.; Van Grieken, R.E.: Laser microprobe mass analysis (LAMMA) of parathyroid glands from dialysis patients; in Brätter, Schramel, Trace element-analytical chemistry in medicine and biology, vol. 4, pp. 443–450 (de Gruyter, Berlin 1987).
34 Ellis, H.A.; McCarthy, J.H.; Herrington, J.: Bone aluminum in haemodialysed patients and in rats injected with aluminum chloride: relationship to impaired bone mineralization. J. clin. Path. *32:* 832–844 (1979).
35 Ellis, H.A.: Personal communication.
36 Terao, N.; Yamamoto, K.; Nakamura, R.; Honma, S.; Asano, Y.; Hosoda, S.: Suppression of parathyroid hormone (PTH) secretion by iron (Abstract). Kidney int. *29:* 173 (1986).
37 Lorenzo, V.; Rodriguez, M.; Felsenfeld, A.; Llach, F.: Aluminum (Al) toxicity in parathyroidectomized rats resembles aplastic bone disease (Abstract). Kidney int. *29:* 164 (1986).
38 Goodman, W.G.: Short-term aluminum administration in the rat: reductions in bone formation without osteomalacia. J. Lab. clin. Med. *103:* 749–757 (1984).

Marc E. De Broe, MD, PhD, Department of Nephrology-Hypertension,
University Hospital Antwerp, Wilrijkstraat 10, B–2520 Antwerp-Edegem (Belgium)

Contr. Nephrol., vol. 64, pp. 144–150 (Karger, Basel 1988)

Methodological Aspects of Trace Element Determination

John Savory [a, b], *Michael R. Wills* [a, c]

Departments of [a]Pathology, [b]Biochemistry and [c]Internal Medicine University of Virginia Medical Center, Charlottesville, Va., USA

Evaluation of aluminum toxicity is to a large extent dependent on the capability of performing accurate measurements of aluminum in biological materials. This analysis is extremely challenging since the analyst is faced with the problem of detecting aluminum in the parts per billion range. Important considerations are specimen collection, storage, processing, and final analysis without outside contamination. The abundance of aluminum in the environment complicates efforts to provide a contamination-free system.

Many problems have existed with analytical methods for the measurement of aluminum. In a review, Cornelis and Schutyser [1984] selected data that had been published since 1974 on aluminum in serum (or plasma) and urine of healthy individuals. Nineteen studies reported mean normal values ranging from 2.1 to 42 μg/l. The other reports they listed had normal mean concentrations ranging from 72 to 1460 μg/l. Obviously, interferences were present in these latter methods. This literature survey also provided normal whole blood mean aluminum concentrations which ranged from 1.6 to 22.1 μg/l (3 studies), and urine levels which varied from 4.7 to 1,700 μg/l (7 studies). In the absence of carefully controlled interlaboratory surveys with well-characterized quality control materials, each laboratory's normal range can be used to provide a useful indicator of interlaboratory variability. The wide variability reported by Cornelis and Schutyser [1984] is undoubtedly due to major analytical problems rather than to biological variation.

Recently, analytical methods for aluminum have been improved considerably. At least two excellent interlaboratory surveys have been introduced to aid the analyst by providing materials which are analyzed in several laboratories.[1] The recognition that normal human serum (or plasma) aluminum levels are less than 10 µg/l has helped many laboratories to identify contamination and analytical problems.

Contamination is a major problem in aluminum analyses. All items utilized during collection, transport, and assay should be checked for unwanted aluminum contribution to the procedure. Only by taking these stringent precautions will one be able to produce results with confidence in their accuracy.

Analytical Methods

Atomic Absorption Spectrometry

Atomic absorption spectrometry (AAS) has been widely used to analyze biological materials for aluminum content. Flame techniques, even with the hotter nitrous oxide acetylene flame, do not perform as well as the flameless methods. The greatest degree of success with any technique for the determination of aluminum in biological specimens has been with electrothermal atomic absorption spectrometry (EAAS). In this technique the sample is placed in a graphite tube mounted in the light path of the spectrophotometer. The source of the light is a hollow cathode lamp that contains the metal being analyzed and emits characteristic wavelengths. First the graphite tube is heated with direct current to dry the sample at a low temperature, then the sample is ashed to destroy organic matter and burn off inorganic species that may interfere, and finally the temperature is quickly raised and the metal under analysis vaporizes and absorbs the light being passed through the graphite tube. Advantages of the graphite furnace include: (1) sample pretreatment can usually be eliminated; (2) sample requirements are small (2–100 µl); (3) graphite furnaces are capable of attaining the high temperature needed to form ground state atoms, and (4) the atoms stay in the light path for a relatively long time which results in increased sensitivity.

Problems encountered in the electrothermal atomic absorption determination of aluminum, include difficulties with untreated graphite tubes [Alderman and Gitelman, 1980], matrix interferences [Alderman and Gitelman, 1980; Gorsky and Dietz, 1978], and standardization procedures [Gorsky and Dietz, 1978].

Recently, Bettinelli et al. [1985] have made a thorough evaluation of the direct measurement of aluminum in serum using EAAS with the stabilized temperature furnace and the L'vov platform or stabilized temperature platform furnace. Their recommendation was

[1] O. Guillard, Laboratoire de Biochimie, Hôpital Jean Bernard, B.P. 577, F-86021 Poitiers Cedex, France; A. Taylor, The Robens Institute, University of Surrey, Guildford, Surrey GU2 5XH, England.

to use pyrolytically coated graphite tubes and to keep the method as simple as possible with minimal sample pre-treatment. Most EAAS procedures have been developed on Perkin-Elmer instruments with the use of auto-sampling to improve precision. Pyrolytically coated graphite tubes are recommended together with a pyrolytic graphite platform. These tubes minimize reactions between the analyte and the graphite tube. Argon is preferred over nitrogen as the purge gas since argon produces a larger and less variable signal. Furnace signals occur rapidly and in order to follow the absorbance profile, it is necessary to incorporate fast digital electronics into the instrument. The absorbance signal must be integrated to provide accurate results; peak absorbance signals will produce inaccurate quantitative measurements. Measurements also require elimination of matrix effects between samples and standards, usually accomplished by some type of background correction. The Zeeman correction system [Slavin, personal communication] provides the most sensitive and reliable results.

Matrix modifiers are an important part of the modern atomic absorption technology. Most analytes require a matrix modifier in order to stabilize the analyte until thermal conditions have become constant. For aluminum, magnesium nitrate permits the use of high char temperatures without losses. The procedure recommended by the present authors is based on that recommended by Bettinelli et al. [1985] and uses a stabilized temperature platform with pyrolytically coated graphite tubes. In this procedure serum analysis standards and serum samples are diluted with an equal volume of an aqueous solution containing $Mg(NO_3)_2$ (2 g/l). The autosampler is programmed to deliver a 10-μl aliquot of the sample or standard onto the platform for final analysis. Aqueous standards are prepared fresh daily. Linearity ranges are from 0 to 125 μg/l. Concentration values are calculated using linear least square regression.

Leung and Henderson [1982] prepare their standards using a serum pool containing a minimal amount of endogenous aluminum, whereas Bettinelli et al. [1985] recommend aqueous standards. The use of simple aqueous standards is to be preferred provided that there are no matrix effects from serum.

The present authors have developed a procedure which minimizes matrix effects by protein precipitation [Brown et al., 1984]. This technique was originally developed for serum nickel [Sunderman et al., 1984] and markedly reduces matrix effects in the final atomic spectroscopic analysis. The sensitivity of all of these serum methods extends down to approximately 1 μg/l.

For urine analysis sample aliquots are diluted 1:1 with distilled water before application to the EAAS stabilized temperature platform [Leung and Henderson, 1982]. Fecal analysis requires considerably more complicated preparation than serum or urine analysis. The procedure developed in the authors' laboratory [Brown et al., 1986] involves sample homogenization, ashing at 550 °C in a muffle furnace, and analysis by EAAS.

Solid tissues must be homogenized, dried, ashed, and/or dissolved to produce a liquid sample prior to EAAS analysis. Bone samples for analysis are washed free of marrow by a strong stream of distilled water. Any existing fat or muscle is scraped with an aluminum-free scalpel. The bone sample can then be processed by any of several methods to obtain a solution for injection into the graphite furnace.

Soft tissue samples, such as brain, liver, or muscle, must be homogenized before processing, and this can be accomplished easily by pummelling the tissue in a 'Stomacher' blender (Fisher Scientific). The homogenate can then be processed as described for the fecal samples. Tissue samples can also be processed by an ethylenediaminetetraacetic acid (EDTA) extraction method. Brain samples are dried and ashed, then extracted with EDTA. Muscle samples are dried, ground, extracted in solvents, redried, and extracted in EDTA [LeGendre and Alfrey, 1976]. Tissue samples can be processed (after homogenization) with hot nitric acid in a Kjeldahl flask.

Other Techniques

Aluminum can be measured using gravimetric, titrimetric, photometric and fluorimetric methods, but interferences from other metals and contamination of reagents with aluminum are major probelms with these techniques. All of these methods require that the aluminum is isolated from interfering metals. Because of poor sensitivity and various other problems, chemical and physicochemical methods are unsuitable for aluminum determinations in serum and other biological materials.

X-Ray fluorescence is another technique, but although this method is very specific and capable of measuring aluminum, it does not appear to be sensitive enough to detect the trace levels of aluminum in serum [Sorenson et al., 1974]. However, one form of X-ray emission spectrography, that of electron probe X-ray microanalysis, has been used effectively to localize aluminum in both bone and brain tissues [Smith and McClure, 1982].

Several investigators have used neutron activation analysis (NAA) to determine the aluminum content of biological specimens both with and without some chemical processing. NAA involves the bombardment of a sample with neutrons and measurement of the radioactivity induced by nuclear reactions. No chemical processing is required. Upon activiation ^{27}Al (100% isotopic abundance) forms the radioactive ^{28}Al nuclide by an (n, γ) reaction. There are attractive features in this technique such as excellent sensitivity with relative independence from matrix effects and interferences. Also, there is relative freedom from contamination since the sample is analyzed with minimal handling. One major problem is the need to correct for fast neutron reactions on phosphorus which also produce ^{28}Al. A detailed description of the technique from the analysis of brain tissue has been reported recently [Ehman et al., 1983].

Other NAA methods involve the chemical separation of interfering ions [Blotcky et al., 1976]. In these the short half-life (2.24 min) of ^{28}Al makes post-irradiation separations a problem.

Aluminum can be measured by emission spectrometry using an argon plasma. Conventional flame emission techniques do not provide sufficient sensitivity for the analysis of biological specimens. Inductively coupled plasma (ICP) methods have been reported for the measurement of aluminum in biological materials [Lichte et al., 1980; Schramel et al., 1980; Mauras and Allain, 1985] and could be an excellent alternative to EAAS for those laboratories possessing the appropriate instrumentation. A major problem with the argon-plasma technique is the intense and broad emission of calcium which increases the aluminum background and can raise the detection limits. Correction for this problem is hampered because the effect of calcium varies with its concentration. However, the sensitivity of ICP methods probably is comparable to EAAS and is reported to be about 2 μg/l [Lichte et al., 1980], although relatively poor sensitivity (10 μg/l) also has been reported [Schramel et al., 1980]. A recent report [Mauras and Allain, 1985] describes the application of ICP to the analysis of blood, dialysis fluid and water. The detection limit of the method is 0.3 μg/l in pure solution with a large range of linearity and excellent precision. This technique offers some potential as a reference method for aluminum measurements. One drawback to ICP, however, is the relatively high cost of the instrumentation.

Standard Reference Materials and Quality Assurance

The availability of standard reference materials is of considerable importance in the application of reliable assays for aluminum. Aqueous standards are available from commercial sources one being Fisher Scientific Company (Fair Lawn, N.J.). This material is in acid solution and is a stock solution of 1 g/l. Dilution to working standards must incorporate an intermediate standard if accuracy is to be maintained. The National Bureau of Standards (Washington, D.C.) also provides an aqueous standard solution which contains 10.00 mg/l of aluminum in a 10% acid medium. Bovine serum (NBS-RM-8419) with a quoted aluminum concentration also is available from the National Bureau of Standards.

Quality assurance in aluminum measurements follows established principles used in clinical chemistry. Internal quality control materials usually pooled plasma, at two or three different concentrations should be used daily. The concentrations of aluminum in these materials should coincide with various decision levels. Interlaboratory control materials also are an important part of any quality assurance scheme. Presently, such a process is conducted at the Robens Institute. The institute mails 3 water and 3 serum samples monthly to laboratories for aluminum analysis. The data and statistical information of the participating laboratories is then made available for review.

One material for monitoring the performance of aluminum assays is plasma from normal individuals. Such a pool would have an aluminum concentration of approximately 5 μg/l and is useful for examining the recovery of aluminum added to normal serum.

References

Alderman, F.R.; Gitelman, H.J.: Improved electrothermal determination of aluminum in serum by atomic absorption spectroscopy. Clin. Chem. *26:* 258–260 (1980).

Bettinelli, M.; Baroni, U.; Fontana, F.; Poisetti, P.: Evaluation of the L'vov platform and matrix modification for the determination of aluminum in serum. Analyst *110:* 19–22 (1985).

Blotcky, A.J.; Hobson, D.; Leffler, J.A.; Rack, E.P.; Recker, R.R.: Determination of trace aluminum in urine by neutron activation analysis. Analyt. Chem. *48:* 1084–1088 (1976).

Brown, S.; Bertholf, R.L.; Wills, M.R.; Savory, J.: Electrothermal atomic absorption spectrometric determination of aluminum in serum with a new technique for protein precipitation. Clin. Chem. *30:* 1216–1218 (1984).

Brown, S.; Mendoza, N.; Bertholf, R.L.; Ross, R.; Wills, M.R.; Savory, J.: Absorption of aluminum from aceglutamide aluminum in healthy adult males. Res. Commun. chem. Pathol. Pharmacol. *53:* 105–115 (1986).

Cornelis, R.; Schutyser, P.: Analytical problems related to Al-determinations in body fluids, water and dialysate; in Quellhorst, Finke, Fuchs, Trace elements in renal insufficiency, pp. 1–11 (Karger, Basel 1984).

Ehman, W.D.; Hossain, T.I.M.; Goodin, D.T.; Markesbery, W.B.: Neutron activation analysis for aluminum in brain tissue; in Wills, Savory, Aluminum analysis in biological materials, pp. 23–33 (University of Virginia Press, Charlottesville 1983).

Gorsky, J.E.; Dietz, A.A.: Determination of aluminum in biological samples by atomic absorption spectrophotometry with a graphite furnace. Clin. Chem. *24:* 1485–1490 (1978).

LeGendre, G.R.; Alfrey, A.C.: Measuring picogram amounts of aluminum in biological tissue by flameless atomic absorption analysis of a chelate. Clin. Chem. *22:* 53–56 (1976).

Leung, F.Y.; Henderson, A.R.: Improved determination of aluminum in serum and urine with use of a stabilized temperature platform furnace. Clin. Chem. *28:* 2139–2143 (1982).

Lichte, F.E.; Hopper, S.; Osborn, T.W.: Determination of silicon and aluminum in biological matrices by inductively coupled plasma emission spectrometry. Analyt. Chem. *52:* 120–124 (1980).

Mauras, Y.; Allain, P.: Automatic determination of aluminum in biological samples by inductively coupled plasma emission spectrometry. Analyt. Chem. *57:* 1706–1709 (1985).

Schramel, P.; Wolf, A.; Klose, B.J.: Direktbestimmung von Aluminium in Serumproben mittels Inductively Coupled Plasma (ICP) – Emissions-Spektralanalyse. J. Clin. Chem. Biochem. *18:* 591–593 (1980).

Smith, P.S.; McClure, J.: Localization of aluminium by histochemical and electron probe x-ray microanalytical techniques in bone tissue of cases of renal osteodystrophy. J. clin. Path. *35:* 1283–1293 (1982).

Sorensen, J.R.J.; Campbell, I.R.; Tepper, L.B.; Lingg, R.D.: Aluminium in the environment and human health. Environ. Health Perspect. *8:* 3–95 (1974).

Sunderman, F.W.; Crisostomo, C.; Reid, M.C.; Hopfer, S.M.; Nomoto, S.: Rapid analysis of nickel in serum and whole blood by electrothermal atomic absorption spectrophotometry. Ann. clin. Lab. Sci. *14:* 232–241 (1984).

John Savory, PhD, Departments of Pathology, Biochemistry and Internal Medicine, University of Virginia Medical Center, Charlottesville, VA 22908 (USA)

Contr. Nephrol., vol. 64, pp. 151–159 (Karger, Basel 1988)

Treatment and Histological Healing of Aluminum-Related Osteomalacia[1]

B.F. Boyce[a], *M.Z. Mocan*[b], *J. Byars*[a], *B.J.R. Junor*[b,2]

[a]University Department of Pathology, Royal Infirmary, Glasgow, and
[b]Renal Unit, Western Infirmary, Glasgow, Scotland

Al-related bone disease is characterized by the presence along the CF (osteoid/calcified bone interface) of Al which appears to cause a mineralization defect by blocking the uptake of calcium into bone [1, 2]. Osteoid seams may [3] or may not [4] be thickened and osteoclastic bone resorption and active bone formation are typically suppressed. Al lines may also be seen within fully calcified bone suggesting that some of the osteoid seams may become mineralized despite the presence of Al along the calcification front. Hypercalcemia and low plasma parathyroid hormone levels are common [5] and the tendency for hypercalcemia to develop is promoted by vitamin D therapy and the addition of calcium to the dialysis fluid.

Al bone disease may be prevented by avoiding the use of Al-containing phosphate binders and of dialysis water contaminated by Al [6] and established cases may be treated with weekly infusions of desferrioxamine (DFO) [7] combined with these preventive measures. DFO treatment appears to facilitate the reemergence of secondary hyperparathyroidism [8, 9] which is accompanied by a reduction in the amount of Al staining within bone. Secondary hyperparathyroidism might also be promoted by lowering the level of calcium in the dialysis fluid and so reversing the hypercalcemia which may contribute to parathyroid gland suppression [5].

[1] This work was partly supported by a grant to B.F. Boyce from the Scottish Hospital Endowments Research Trust.

[2] We thank Mrs. Thelma Barrios for typing the manuscript.

In this paper we review the bone biopsy finding of 14 patients with Al bone disease. Six had been given no active therapy, 5 were given DFO and in 3 the level of calcium in the dialysis fluid was lowered.

No Active Therapy Group

This group consisted of 6 patients (5 males, one female, aged 20–40 years) with established chronic renal failure. Males had received regular hemodialysis treatment for between 27 months and 9 years with water containing low levels of Al (generally less than 1,5 μmol/l). The female had been treated with continuous ambulatory peritoneal dialysis for 32 months before bone biopsy. All patients had been taking Al-containing phosphate binders up to the time of bone biopsy.

All patients had extensive, thickened osteoid seams (osteoid surfaces >60%, maximum number of birefringent osteoid lamellae ⩾7) and extensive Al staining along the CF (>67%). Al lines were also present along cement lines in many of the bone trabeculae. A prominent feature in all of the specimens was the presence of an unusual form of patchy calcification within the thickened osteoid seams that was lighter staining than the underlying fully calcified bone. This incomplete form of calcification occupied between 13 and 69% of the osteoid area in undecalcified sections stained with 1% aqueous toluidine blue. Its distribution within the thickened trabeculae varied from predominantly periosteocytic in a few areas (fig. 1) to a more diffuse pattern (fig. 2) in which calcification was more intense. The calcification was confirmed by means of Von Kossa staining, and by energy dispersive electron probe X-ray microanalysis of ultrathin sections cut from the blocks using a technique which we have described previously [1].

These observations indicate that thickened osteoid seams can calcify by means of an unusual form of calcification that may begin around osteoid osteocytes and subsequently spread diffusely within the osteoid down to the Al line at the underlying osteoid/calcified bone interface. Complete mineralization of this partially calcified osteoid could account for the presence of Al lines within calcified bone [10].

The presence of patchy calcification within osteoid in the absence of active treatment for Al bone disease raises the possibility that Al toxicity may be intermittent in patients whose main source of Al is phosphate binders. Variation from time to time in Al absorption from the gastrointestinal tract could result in fluctuations in serum, and thus bone, levels of

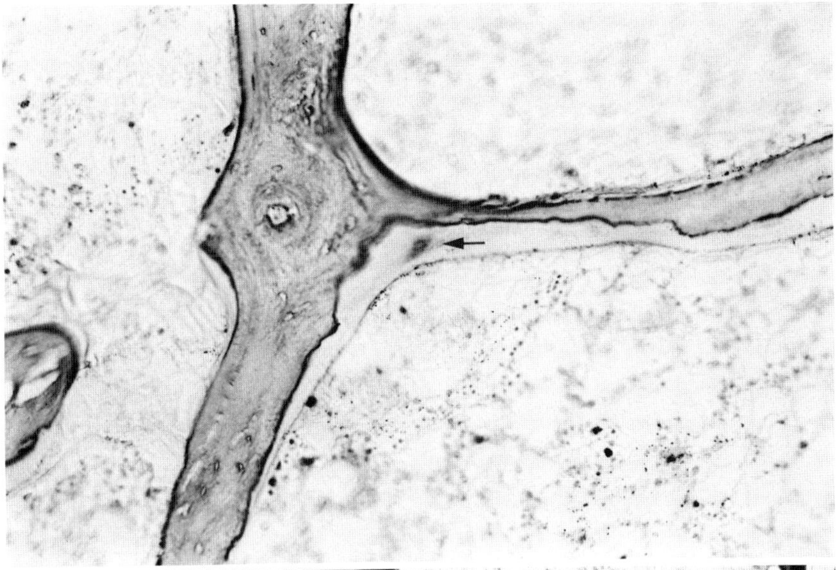

1

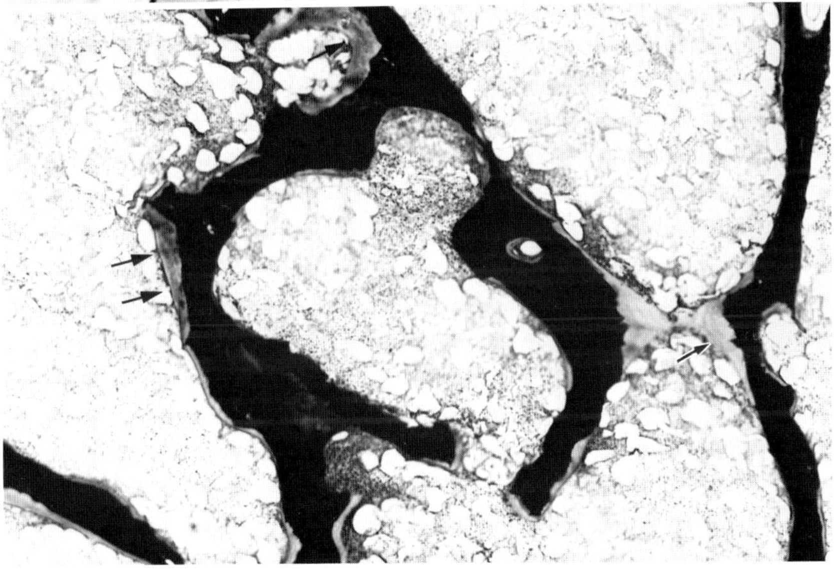

2

Fig. 1. Undecalcified section of bone stained for Al. A small focus of calcification (arrowed) is present around an osteocyte within a thickened osteoid seam. Al appears as a dark line between pale staining osteoid and the darker staining calcified bone. ×137.

Fig. 2. Undecalcified section of bone stained with 1% aqueous toluidine blue. More extensive calcification (large arrows) is present within some thickened osteoid seams but is absent in others (small arrow). ×55.

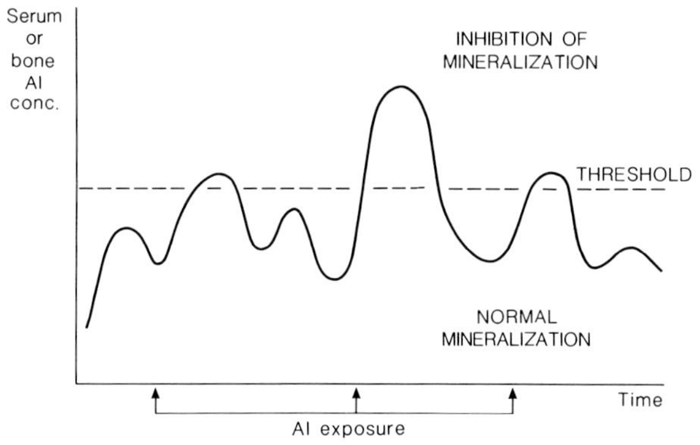

Fig. 3. Inhibition of mineralization threshold. Increased Al exposure could be due to increased Al absorption from the gut or to rises in dialysate Al levels.

Al around a threshold above which mineralization would be inhibited. As illustrated in figure 3 episodes of increased exposure to Al might result in this threshold being crossed with subsequent deposition of Al and inhibition of mineralization. Histological healing with calcification of osteoid as described above could occur during those periods when the level of Al in the circulating bone fluid is below this proposed threshold.

DFO-Treated Group

This group consisted of 5 patients (2 males, 3 females, aged 29–53 years) who had received regular hemodialysis with low Al water during 3–8 years. All had been taking phosphate binders which were stopped before DFO treatment was started. Transiliac bone biopsies were taken before and at the end of 5–8 months of treatment with weekly infusions of 6 g DFO. Therapy was continued for a further 12 months in 2 patients at which time a third bone biopsy was taken and treatment was stopped.

Before the start of therapy 4 patients had extensive, thickened osteoid seams (osteoid surface >60%, maximum number of birefringent lamellae of osteoid ⩾6) with extensive Al staining along the calcification front (>66%), typical of Al-related osteomalacia. In one patient the extensive osteoid seams were mainly of normal thickness; Al was present along 77% of the osteoid and thus the picture was more typical of 'aplas-

Table I. Bone histomorphometry in 5 patients treated with DFO

	Al staining % osteoid surface range (mean)	Resorption surface % trabecular surface range (mean)	Number of osteoclasts per mm^2 range (mean)
Group 1 (n=4)			
Before DFO	66–94 (78.3)	2.6–18 (7.7)	0.3–1.8 (1.0)
After DFO	1.6–55 (26.6)	17–30 (24.1)	1.0–6.5 (3.25)
Group 2 (n=1)			
Before DFO	77.0	1.7	0.24
After DFO	68.4	5.9	0.24
Normal value	0	<7.3	<0.28

tic' Al bone disease. Osteoclastic bone resorption was within the normal range in 4 patients but was high (resorption surface 17.9%) in one of the osteomalacic patients.

The changes in bone histomorphometry are listed in table I. In 4 of the patients (group 1) there was a significant reduction in the extent of Al staining along osteoid seams. However, the response was variable and in one case Al was still present along 55% of the calcification front after 5 months of treatment. Osteoclastic bone resorption rose significantly in 3 of these patients and remained elevated in the fourth.

The extent of Al staining and the number of osteoclasts did not change significantly in the 5th patient (group 2) and the resorption surface rose slightly but remained within the normal range. DFO was continued in this patient for a further 12 months but this still did not result in a reduction in Al staining or in an increase in osteoclastic bone resorption. DFO therapy was also continued for 12 months in the patient who had the most pronounced response to therapy (Al staining had fallen from 72.9 to 1.6% after 6 months). No Al staining was seen in her 3rd biopsy specimen.

These findings indicate that as in previous studies [8, 9] DFO therapy is associated with a reduction in the extent of Al staining along the CF in most cases of Al-related osteomalacia and that this is accompanied by the reemergence of secondary hyperparathyroidism. However, the bone response is variable [8, 9]. The reason why some patients fail to respond remains unknown but it is possible that 'aplastic' Al bone disease is more resistant to therapy.

A point of interest was that Al lines were still present within calcified bone in 3 of 4 patients in group 1 and in the patient in group 2 at the end of therapy. Abundant hemosiderin pigment was also seen in macrophages within bone marrow before and after therapy. Thus, DFO does not appear to remove Al or iron that is fixed in bone or cells by some form of elution. Rather, Al appears to be removed by osteoclasts which resorb the Al-rich bone.

Low Calcium Dialysate Treated Group

This group consisted of 3 patients (aged 20–60 years, 2 males, one female) who had received regular hemodialysis for 2–3 years. Only one had been exposed to a high water Al level which was subsequently controlled by reverse osmosis and de-ionization. All three had taken Al-containing phosphate binders since before starting on regular hemodialysis. At the time of the initial bone biopsy all three had hypercalcemia (serum calcium >2.7 mmol/l [normal range 2.2–2.6]) which had been present for 12–18 months before the calcium level of the dialysis fluid was lowered from 1.5 to 1.34 mmol/l. Low calcium dialysate was used for 12–29 months during which time 2 or 3 additional bone biopsies were taken. Phosphate binders had been stopped before the use of a low calcium dialysate which was introduced in an attempt to lower the serum calcium and treat the Al bone disease.

Two of the patients had typical Al-related osteomalacia (osteoid surface >66%, maximum number of birefringent lamellae of osteoid ⩾12, Al staining along >85% of the osteoid/calcified bone interface and normal bone resorption) which had remained unaltered since a previous biopsy taken 8 months earlier. In the third patient the osteoid surface was at the upper end of the normal range (25%) but Al staining was present along 90% of its interface with calcified bone and bone resorption was normal. Osteoid was of normal thickness in trabecular bone but was thickened in the cortex (8 birefringent lamellae). Thus, the appearances in this case were of minimal Al bone disease despite high serum levels of calcium and Al.

Al staining had disappeared from the CF in biopsy specimens taken from patients 1 and 2 after 23 and 5 months of treatment, respectively. In both cases this was accompanied by increased osteoclastic bone resorption and the virtual disappearance of Al from within fully calcified bone. In a biopsy specimen taken from patient 1 after 9 months of treatment extensive Al staining was still present along 75% of the osteoid although by this time osteoclastic bone resorption had reemerged (resorption surface 12,9%). Hypercalcemia persisted in both patients during the first 9 months of treatment after which values returned to within the normal range.

In contrast to the above findings the histological response in patient 3 was less pronounced. Despite a fall in the extent of Al along osteoid surfaces, the extent of osteoid was much higher in the specimen taken after 13 months of treatment so that Al staining overall was more extensive. This difference in extent of osteoid is more likely to be due to site to site variation within the ilium before treatment rather than to a real increase in osteoid surfaces during treatment since osteoclastic bone resorption remained within the normal range and had not been increased in a specimen taken after 9 months (resorption surface=2.9%). Nevertheless, serum calcium levels returned to the normal range in this patient after 7 months of treatment.

The bone biopsy findings in this small study indicate that the use of a low calcium level in the dialysis fluid and withdrawal of phosphate binders may result in improvement and eventual healing of Al-related osteomalacia. Histological improvement appears to be mediated through the reemergence of secondary hyperparathyroidism and osteoclastic bone resorption. However, the rate of response is variable and, as is seen in patient 3, may not be apparent even after 13 months. The explanation for this failure of delay in response is not evident.

Comment

Variable degrees of improvement of Al-related osteomalacia appear to accompany the use of DFO or low calcium dialysate. Although DFO therapy is currently considered by some to be the treatment of choice, it is expensive and may have troublesome side effects [11, 12]. In contrast, a low calcium dialysate is inexpensive and is unlikely to have any side effects providing hypocalcemia is avoided. Both forms of therapy faciliate healing of Al-related osteomalacia by promoting the development of secondary hyperparathyroidism and treatment may need to be continued for several months before a histological cure is achieved. As a result of the widespread use of water treatment methods such as reverse osmosis to maintain low dialysate Al levels, phosphate binders are now becoming the main cause of Al bone disease and are associated with less acute clinical problems. We suggest that a low calcium dialysate might be as effective as DFO therapy in the treatment of Al bone disease particularly in patients in whom a rapid clinical response is not mandatory. We also suggest that a combination of DFO and a low calcium dialysate might be worth a trial in those patients who do not respond to either therapy alone.

References

1 Boyce, B.F.; Elder, H.Y.; Fell, G.S.; Nicholson, W.A.P.; Smith, G.D.; Dempster, D.W.; Gray, C.C.; Boyle, I.T.: Quantitation and localisation of aluminium in human cancellous bone in renal osteodystrophy. Scann. electron Microsc. *III:* 329–337 (1981).

2 Cournot-Witmer, G.; Zingraff, J.; Plachot, J.J.; Escaig, F.; Lefèvre, R.; Boumati, P.; Bourdeau, A.; Garabédian, M.; Galle, P.; Bourdon, R.; Drüeke, T.; Balsan, S.: Aluminum localization in bone from hemodialyzed patients: relationship to matrix mineralization. Kidney int. *20:* 375–385 (1981).

3 Ellis, H.A.; McCarthy, J.H.; Herrington, J.: Bone aluminium in haemodialysed patients and in rats injected with aluminium chloride: relationship to impaired bone mineralisation. J. clin. Path. *32:* 832–844 (1979).

4 Charhon, S.A.; Chavassieux, P.M.; Chapuy, M.C.; Boivin, G.Y.; Meunier, P.J.: Low bone formation with and without histologic appearance of osteomalacia in dialyzed patients with aluminum intoxication. J. Lab. clin. Med. *106:* 123–131 (1985).

5 Boyce, B.F.; Fell, G.S.; Elder, H.Y.; Junor, B.J.; Elliot, H.L.; Beastall, G.; Fogelman, I.; Boyle, I.T.: Hypercalcaemic osteomalacia due to aluminium toxicity. Lancet *ii:* 1009–1013 (1982).

6 Kerr, D.N.S.; Ward, M.K.; Arze, R.S.; Ramos, M.; Grekas, D.; Parkinson, I.S.; Ellis, H.A.; Owen, J.P.; Simpson, W.; Dewar, J.; Martin, A.M.; McHugh, M.F.: Aluminum-induced dialysis osteodystrophy: the demise of 'Newcastle bone disease'? Kidney int. *29:* S58–S64 (1986).

7 Ackrill, P.; Day, J.P.; Garstang, F.M.; Hodge, K.C.; Metcalfe, P.J.; Benzo, Z.; Hill, K.; Ralston, A.J.; Ball, J.; Denton, J.: Treatment of facturing renal osteodystrophy by desferrioxamine. Proc. Eur. Dial. Transpl. Ass. *19:* 203–207 (1982).

8 Ott, S.M.; Andress, D.L.; Nebeker, H.G.; Milliner, D.S.; Maloney, N.A.; Coburn, J.W.; Sherrard, D.J.: Changes in bone histology after treatment with desferrioxamine. Kidney int. *29:* S108–S113 (1986).

9 Charhon, S.A.; Chavassieux, P.M.; Chapuy, M.C.; Accominotti, M.; Traeger, J.; Meunier, P.J.: Traitment par la desferrioxamine de l'osteomalacie par intoxication à l'aluminium. Presse méd. (in press).

10 Boyce, B.F.; Junor, B.J.R.; Elder, H.Y.; McWilliams, S.: Histological healing of aluminium-related osteomalacia leads to multiple Al lines in bone. Bone *6:* 227 (1986).

11 Garrett, P.; O'Connor, M.; Walker, J.F.; Donohoe, J.F.; Carmody, M.; Dervan, P.: Desferrioxamine chelation therapy of aluminium intoxication syndrome: the need

for a controlled study; in Taylor, Aluminium and other trace elements in renal disease, pp. 200–202 (Baillière, Tindall, London 1986).
12 Ackrill, P.: Aluminium removal by desferrioxamine: clinical practice; in Taylor, Aluminium and other trace elements in renal disease, pp. 139–199 (Baillière, Tindall, London 1986).

B.F. Boyce, MD, University Department of Pathology, Royal Infirmary, Glasgow, G4 OSF (Scotland)

Contr. Nephrol., vol. 64, pp. 160–169 (Karger, Basel 1988)

Deferoxamine for the Diagnosis and Treatment of Aluminum-Associated Osteodystrophy

S. Charhon [a], *S. El Yafi* [b], *M. Gaillard* [c], *M. Beruard* [d], *G. Janin* [b], *M. Accominotti* [e], *J. Traeger* [f], *P.J. Meunier* [a, 1]

[a]INSERM U. 234, Lyon; [b]Unité d'Hémodialyse, Hôpital de Macon; [c]Unité d'Hémodialyse, Hôpital du Puy; [d]Unité d'Hémodialyse, Clinique du Tonkin, Villeurbanne; [e]Laboratoire de Biochimie D, and [f]Clinique de Néphrologie, Hôpital E. Herriot, Lyon, France

Since the first report that deferoxamine (DFO) was able to improve the clinical condition of aluminum (Al)-intoxicated patients by chelating Al [1], this compound has been widely used in both diagnosis and treatment of Al overload. DFO, a well-known chelating agent of iron, induces a rise in plasma Al of uremic patients on chronic maintenance dialysis. Both the ultrafiltrable and protein-bound Al concentrations increase after a DFO infusion. Serial measurements of plasma Al following a single i.v. DFO infusion administered after a dialysis have shown that the maximal increase in plasma Al was reached after 18–24 h [11]. Afterwards, plasma Al remains stable for at least 24 h. By analogy with the standardized DFO infusion which has been used to assess the tissue iron stores by measuring the urinary excretion of iron, a standardized DFO test has been proposed to assess the tissue Al loading in dialysis patients by measuring the increase in plasma Al induced by DFO [11, 13]. Indeed, it has been shown that the measurement of only the plasma Al was not reliable for the diagnosis of Al loading and Al bone disease. In particular, if a high plasma Al value (over 100 µg/l) indicates a high probability of Al

[1] The authors are endebted to Dr F. Ducret for providing us with biopsy material from patients under its care and Drs G. Boivin, P. Chavassieux, M. Parisien and M.C. Chapuy for their contribution to the histological and biochemical studies.

deposits in bone, a lower value does not exclude the presence of Al bone deposits, or even of Al-related osteomalacia [3]. The fact that plasma Al decreases rapidly after discontinuation of Al exposure suggests that plasma Al reflects more the recent exposure to Al rather than the tissue Al stores.

DFO Infusion Test for the Diagnosis of Aluminum-Associated Bone Disease

To validate the DFO infusion test as a noninvasive method for assessing the tissue Al stores, several studies have analyzed the correlations between the DFO-induced increase in plasma Al (Δ Al) and the bone Al content (BAC) measured by atomic absorption spectrometry (AAS) [7, 11, 12], the extent of stainable Al deposits on trabecular bone surfaces [8, 11, 12] or the cumulative dose of Al hydroxide ingested [7]. The baseline plasma Al and the Δ Al values in patients with Al loading (as defined by stainable bone Al) and in patients without Al loading (negative staining) have also been compared [2, 10]. In these studies, i.v. DFO was administered at various doses ranging from 2 to 6 g either during the second part of a dialysis [2, 7, 14], or immediately after a dialysis to minimize the loss of free DFO during dialysis [8, 11, 12]. In contrast, in the study of Malluche et al. [10], DFO was infused during the first 2 h of a dialysis and measurements of plasma Al were performed 5 h after the beginning of the DFO infusion instead of approximately 48 h in the other studies. Significant correlations were found between Δ Al and both stainable bone Al [11, 12] and BAC [7, 11, 12]. The latter two parameters were more closely correlated with Δ Al than with baseline plasma Al [7, 11, 12] suggesting that the DFO test might be useful for assessing the Al loading. However, Hodsman et al. [8] did not find any significant difference in plasma Al and ΔAl between patients with Al-related osteomalacia (Al-OM) and those with osteitis fibrosa (OF) having negative Al staining. Similarly, in Malluche's study, patients having Al accumulation in bone and patients without Al accumulation had similar Δ Al values.

In a study designed to determine the value of a single DFO infusion as an indirect method for the assessment of the tissue Al stores and to analyze the DFO dose effect on Al mobilization, 25 dialysis patients underwent 2 DFO infusions (40 mg/kg and 100 mg/kg) and a transiliac bone biopsy.

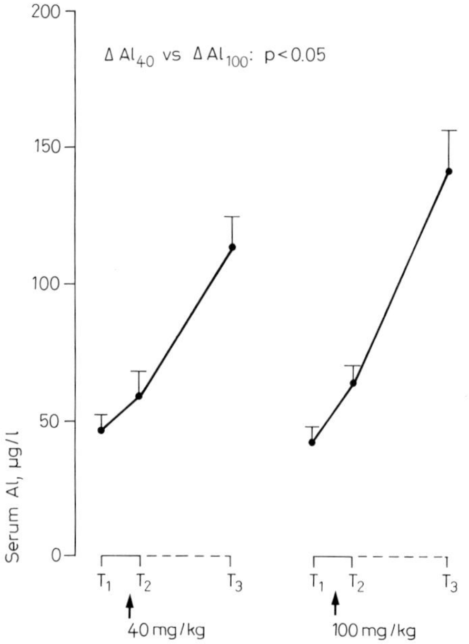

Fig. 1. Mean serum Al values ± SEM before and after a single DFO infusion at 40 mg/kg and 100 mg/kg body weight.

Patients and Methods

We studied 25 adult patients undergoing chronic hemodialysis for at least one year. There were 11 females and 14 males who dialyzed in 3 centers around Lyon. The Al concentration of dialysis water was less than 15 µg/l during the DFO test. Each patient received 2 DFO infusions at an interval of 14 days. The rank of administration of the tests was randomized; 14 patients first received the 100-mg/kg dose and 11 patients first received the 40-mg/kg dose. DFO was administered intravenously during the last 2 h of a dialysis. One week before DFO tests, Al hydroxide gels were discontinued.

Measurements of serum Al were performed by AAS at the beginning (T1), the end (T2) of 'DFO' dialysis, and again 44–48 h later (T3). Transiliac bone biopsies were performed after a double tetracycline labelling allowing a histomorphometric study of bone remodelling and the measurement of the extent of stainable Al on trabecular bone surfaces (AlS) by Aluminon staining using previously published methods [4]. A second core was used for the determination of the bone Al content by AAS.

Table I. Coefficients of linear and power correlations between the bone Al content (BAC) or the Al trabecular surfaces (AlS) and the baseline serum Al (Al_{T1}), the increases in serum Al induced by the 40 mg/kg DFO dose ($\Delta\ Al_{40}$) and the 100-mg/kg dose ($\Delta\ Al_{100}$)

	Al_{T1}	$\Delta\ Al_{40}$	$\Delta\ Al_{100}$
Linear correlation			
BAC	r=0.17	r=0.29	r=0.30
AlS	r=0.09	r=0.17	r=0.20
Power correlation			
BAC	r=0.17	r=0.44*	r=0.49*
AlS	r=0.15	r=0.39	r=0.28

* Significant at $p<0.05$.

Results

Figure 1 shows that baseline serum Al (Al_{T1}) was similar for both tests. Serum Al_{T3} was higher with the 100-mg/kg DFO dose (142.2 ± 15.0 μg/l) than with the 40-mg/kg dose (113.2 ± 10.5 μg/l) but the difference was not significant. In contrast, the increase in serum Al ($\Delta Al = Al_{T3} - Al_{T1}$) was significantly greater with the high dose (99.8 ± 12.7 μg/l) than with the low dose (66.4 ± 9.4 μg/l; $p<0.05$). The BAC was moderately but significantly increased when compared to control values (38.2 ± 5.9 vs 7.8 ± 4.5 μg/g ash weight, $p<0.001$). No patient had Al-OM but nine had stainable Al in bone. No significant linear or power correlation was found between baseline serum Al and BAC (table I). The linear correlation between Δ Al of both tests and BAC was not significant whereas the power correlations were significant ($p<0.05$). Correlation coefficients between Δ Al and AlS were slightly lower.

Discussion

These data show that a 100-mg/kg i.v. DFO infusion induces a Δ Al 50% higher than that induced by a 40-mg/kg dose in patients with moderate Al loading. Both Δ Al values were correlated with the BAC and had similar correlation coefficients. Therefore, the higher dose did not result in any advantage for the assessment of Al loading. As previously reported, the correlation between Δ Al and BAC was closer than between baseline serum Al and BAC thereby confirming that Δ Al is a more

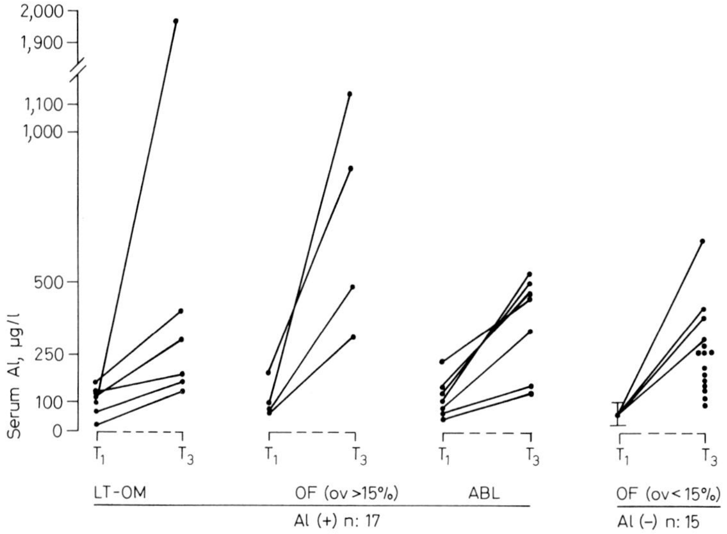

Fig. 2. Increases in serum Al after a 2-gram DFO infusion in patients having various bone lesions with positive Al staining (Al+) or negative Al staining (Al–). LT-OM = Low turnover osteomalacia; OF = osteitis fibrosa, with (OV > 15%) or without osteomalacia (OV < 15%); ABL = aplastic bone lesion; ov = osteoid volume.

accurate index of the tissue Al stores than baseline serum Al. In the study of Nebeker et al. [12] a Δ Al lower than 200 μg/l indicated a low probability of Al bone disease and a value higher than 500 μg/l a high probability of Al bone disease. However, the Δ Al value did not point to the underlying bone lesions in most cases. Our own observations (fig. 2) further support this conclusion since a Δ Al of similar magnitude was observed after a 2-gram DFO infusion in many patients with Al-OM, in others with a mixed bone disease or 'aplastic' bone lesion with positive Al staining, and in some with pure OF and negative Al staining. These findings were not unexpected since it has been shown that patients with OF could have BAC values similar to those with Al-OM [6].

Although the combined measurement of serum iPTH and osteocalcin levels, which we have found to be well correlated with the bone formation rate [4], improves the identification of Al-loaded patients with low bone formation, a bone biopsy after double tetracycline labelling is often necessary for a precise determination of bone lesions.

Table II. Mean values ± SD of histomorphometric parameters before and after DFO treatment in 14 patients with Al-associated bone disease

	Before DFO	After DFO
Al surfaces (N: 0%)	67.7 ± 21.7	26.8 ± 26.4†
Osteoid volume (N: 2.5 ± 1.6%)	19.5 ± 17.0	10.6 ± 6.5†
Osteoid surfaces (N: 13.2 ± 7.1%)	64.4 ± 20.8	53.4 ± 19.2**
Osteoid seam thickness index (N: 18.5 ± 4.1)	28.2 ± 16.0	18.8 ± 6.9**
Osteoclast count (N: 0.20 ± 0.09/mm^2)	0.46 ± 0.39	0.94 ± 1.06*
BFR_{BMU} (µm^3/µm^2/day)	0.008 ± 0.014	0.150 ± 0.185†
BFR_{tissue} (µm^3/µm^2/day)	0.006 ± 0.010	0.084 ± 0.111†

* $p = 0.05$; ** $p < 0.05$; † $p < 0.01$.

DFO for the Treatment of Aluminum-Associated Bone Disease

A rapid improvement in bone pain and fractures as well as in bone histology has been reported subsequent to repeated DFO infusions in dialysis patients with Al-OM [1, 9, 10, 13]. This was achieved with doses ranging from 2 to 6 grams, administered intravenously in 1–3 infusions per week or intraperitoneally.

Herein we report the histological effects on bone of DFO infusions administered once weekly in 14 hemodialysis patients, all having low bone formation rate (BFR), increased osteoid volume and stainable bone Al.

Patients and Methods

There were 9 males and 5 females, aged 34–60, on chronic maintenance hemodialysis three times a week for 14–84 months (mean: 47 ± 21). Nine patients had typical Al-OM with increased thickness index of osteoid seams and five had 'aplastic' bone lesions (ABL) with increased osteoid volume (OV) but normal osteoid seam thickness. DFO was given intravenously at a dose of 6 g in 2 patients, 3 g in 3, 2 g in 8 and 1 g in 1 patient.

Table III. Correlation coefficients between the reduction in OV and other variables in 12 patients treated for 6 ± 1 months with DFO

	Reduction in OV, %
Osteoclast count	
Pre-DFO	r′ = 0.48
Post-DFO	r′ = 0.59*
Serum iPTH	
Pre-DFO	r′ = 0.75**
Post-DFO	r′ = 0.73**
Length of therapy	r′ = 0.05
Total DFO dose administered, mg/kg body weight	r′ = −0.30

* $p < 0.05$; ** $p < 0.01$.

Nine patients continued to receive Al containing gels at various doses throughout the DFO therapy and two during a part of the study. Seven continued to receive oral calcium supplements. Low dose vitamin D supplementation given before the initiation of DFO treatment was also continued in nonhypercalcemic patients. This therapy had been given for at least 8 months before the study but failed to overcome the mineralization defect as demonstrated by the low BFR on bone biopsies taken before DFO treatment. Vitamin D supplements consisted of $25(OH)D_3$ (25 µg/day) in 3 patients, $1\alpha\text{-}(OH)D_3$ (0.25 µg/day) in 1 patient and a combination of $1.25(OH)_2D_3$ and $24,25(OH)_2D_3$ in 3 patients [5].

Bone biopsies were taken before and after 7.1 ± 3.3 months of DFO treatment. Table II lists the histological parameters measured using previously published methods [4]. Serum iPTH levels (C-terminal fragment) were also measured [4].

Results

As shown in table II, there was a significant reduction in Al-covered surfaces (Al-S) and in the osteoid parameters while the bone formation rates at BMU and tissue levels increased significantly (Wilcoxon T test). The osteoclast count (OC) also increased, as well as the osteoblastic osteoid surfaces (data not shown). However, 6 patients showed no change or only mild changes in double-labelled surfaces and OV (poor responders) and 8 showed marked improvements in these parameters (good responders). The weekly time on dialysis, the DFO doses and the length of therapy did not differ significantly between the 2 groups. Poor responders differed from good responders by significantly lower pre- and post-treatment values of OC ($p<0.05$ and $p<0.01$, respectively) and lower post-treatment values of iPTH (7.6 ± 5.5 vs 22.8 ± 11.8 mIU/ml, $p=0.01$, Mann-Whitney U test).

To eliminate length of therapy and DFO doses as variables, we analyzed in the 12 patients who were treated for 5.9 ± 0.9 months, the correlations between the reduction in OV (Δ OV), expressed in percent of initial value, and parameters such as OC, serum iPTH, length of therapy and total DFO dose administered, expressed in mg/kg body weight. Δ OV was correlated with the post-treatment value of OC and iPTH levels (table III). In contrast, no significant correlation was found between Δ OV and length of therapy or DFO doses. After treatment, BFR tissue was significantly correlated with iPTH values ($r'=0.63$, $p<0.05$).

Discussion

These data confirm and extend the results of previous studies showing that the DFO treatment induced a significant reduction in the Al surfaces and in the osteoid excess, and an increase in bone mineralization. This was achieved with moderate DFO doses (25–40 mg/kg) administered only once a week in some patients. However, a heterogeneity in the responses to DFO was noted; some patients (mainly ABL) showing a poor response whereas some others (mainly typical Al-OM) showed a marked improvement in bone histology. Interestingly, the percent reduction in OV was significantly correlated with biochemical and histological evidence of hyperparathyroidism after treatment. In addition, poor responders had lower values of OC before DFO treatment. These observations are consistent with those of Ott et al. [13] who noted a greater increase in bone formation after DFO treatment in patients who had not had a parathyroidectomy and those of Smith and Winney [15] who noted that good responders to dialysis water treatment tended to have higher pretreatment values of iPTH than poor responders. These data are consistent with the concept that PTH might have a facilitating role in bone mineralization. However, there is another possible interpretation according to which the rise in PTH levels results from the improvement of bone mineralization: the removal of Al from bone and the increased calcium accretion into bone might induce a decrease in serum calcium which, in turn, stimulates the PTH secretion.

From these data, it is concluded that DFO is an effective treatment of Al-OM and that presence of hyperparathyroidism is associated with an increased response to DFO.

References

1 Ackrill, P.; Day, J.P.; Garstang, F.M.; Hodge, K.C.; Metcalfe, P.J.; Benzo, Z.; Hill, K.; Ralston, A.J.; Ball, J.; Denton, J.: Treatment of fracturing renal osteodystrophy by desferrioxamine. Proc. Eur. Dial. Transpl. Ass. *19:* 203–207 (1982).

2 Berland, Y.; Charhon, S.A.; Olmer, W.; Meunier, P.J.: Predictive value of Desferrioxamine infusion test for bone aluminum deposits in hemodialyzed patients. Nephron *40:* 433–435 (1985).

3 Charhon, S.A.; Chavassieux, P.M.; Meunier, P.J.; Accominotti, M.: Serum aluminium concentration and aluminium deposits in bone in patients receiving haemodialysis. Br. med. J. *290:* 1613–1614 (1985).

4 Charhon, S.A.; Delmas, P.D.; Malaval, L.; Chavassieux, P.; Arlot, M.; Chapuy, M.C.; Meunier, P.J.: Serum bone gla-protein in renal osteodystrophy: comparison with bone histomorphometry. J. clin. Endocr. Metab. *63:* 892–897 (1986).

5 Chavassieux, P.M.; Charhon, S.A.; Meunier, P.J.: Effects of dihydroxyvitamin D_3 and desferrioxamine therapy on aluminum-related osteomalacia; in Norman, 6th Workshop on Vitamin D, Merano 1985, Vitamin D. A Chemical, biochemical and clinical update, pp. 954–955 (de Gruyter, Berlin 1985).

6 Cournot-Witmer, G.; Zingraff, J.; Plachot, J.J.; Bourdeau, A.; Garabedian, M.; Escaig, F:; Lefevre, R.; Galle, P.; Bourdon, R.; Druecke, T.; Balsan, S.: Aluminum localization in bone from hemodialyzed patients: relationship to matrix mineralization. Kidney int. *20:* 375–385 (1981).

7 Fournier, A.; Fohrer, P.; Leflon, P.; Morinière, P.; Tolani, M.; Lambrey, G.; Demontis, R.; Sebert, J.L.; Van de Vyver, F.; De Broe, M.: The desferrioxamine test predicts bone aluminum burden induced by $Al(OH)_3$ in uraemic patients but not mild histological osteomalacia. Proc. Eur. Dial. Transpl. Ass. *21:* 371–376 (1984).

8 Hodsman, A.B.; Hood, S.A.; Brown, P.; Cordy, P.E.: Do serum aluminum levels reflect underlying skeletal aluminum accumulation and bone histology before or after chelation by Deferoxamine?. J. Lab. clin. Med. *106:* 674–681 (1985).

9 Ihle, B.U.; Buchanan, M.R.C.; Stevens, B.; Becker, G.J.; Kincaid-Smith, P.: The efficacy of various treatment modalities on aluminium-associated bone disease. Proc. Eur. Dial. Transpl. Ass. *19:* 195–200 (1982).

10 Malluche, H.H.; Smith, A.J.; Abreo, K.; Faugere, M.C.: The use of deferoxamine in the management of aluminum accumulation in bone in patients with renal failure. New Engl. J. Med. *311:* 140–144 (1984).

11 Milliner, D.S.; Nebecker, H.G.; Ott, S.M.; Andress, D.L.; Sherrard, D.J.; Alfrey, A.C.; Slatopolsky, E.A.; Coburn, J.W.: Use of the deferoxamine infusion test in the diagnosis of aluminium-related osteodystrophy. Ann. intern. Med. *101:* 775–780 (1984).

12 Nebeker, H.G.; Andress, D.L.; Milliner, D.S.; Ott, S.M.; Alfrey, A.C.; Slatopolsky, E.A.; Sherrard, D.J.; Coburn, J.W.: Indirect methods for the diagnosis of aluminum bone disease: plasma aluminum, the desferrioxamine infusion test, and serum iPTH. Kidney int. *29:* 596–599 (1986).

13 Ott, S.M.; Andress, D.L.; Nebeker, H.G.; Milliner, D.S.; Maloney, N.A.; Coburn, J.W.; Sherrard, D.H.: Changes in bone histology after treatment with desferrioxamine. Kidney int. *29:* S-108–S-113 (1986).

14 Simon, P.; Ang, K.S.; Tanquerel, T.: Surcharge tissulaire en aluminium chez les hémodialysés. Test à la Desferrioxamine. Nouv. Presse Méd. *11:* 209 (1982).
15 Smith, G.D.; Winney, R.J.: Aluminium-related osteomalacia, response to reverse osmosis water treatment; in Taylor, Aluminium and other trace elements in renal disease, Int. Conf. on Aluminium in Renal Disease, Guilford 1985, pp. 98–107 (Baillière, Tindall, London 1986).

S.A. Charhon, MD, INSERM U.234, Faculté Alexis Carrel, 8, rue Paradin, F–69008 Lyon (France)

Subject Index